HOW GREEK SCIENCE
PASSED TO THE
ARABS

HOW GREEK SCIENCE PASSED TO THE ARABS

By

De Lacy O'Leary, D.D.

Lecturer in Bristol University

LONDON

ROUTLEDGE & KEGAN PAUL LTD

BROADWAY HOUSE : 68–74 CARTER LANE, E.C.4

First published 1949
Second impression 1951

Printed in Great Britain by Butler & Tanner Ltd., Frome and London

CONTENTS

PAGE

I. INTRODUCTION I

II. HELLENISM IN ASIA 6
   1. Hellenization of Syria 6
   2. The Frontier Provinces 10
   3. Foundation of Jundi-Shapur . . . 14
   4. Diocletian and Constantine . . . 15

III. THE LEGACY OF GREECE 19
   1. Alexandrian Science 19
   2. Philosophy 21
   3. Greek Mathematicians 29
   4. Greek Medicine 34

IV. CHRISTIANITY AS A HELLENIZING FORCE . . 36
   1. Hellenistic Atmosphere of Christianity . 36
   2. Expansion of Christianity . . . 40
   3. Ecclesiastical Organization . . . 44

V. THE NESTORIANS 47
   1. First School of Nisibis 47
   2. School of Edessa 50
   3. Nestorian Schism 52
   4. Dark Period of the Nestorian Church . 62
   5. The Nestorian Reformation . . . 64

VI. THE MONOPHYSITES 73
   1. Beginning of Monophysitism . . . 73
   2. The Monophysite Schism . . . 75
   3. Persecution of the Monophysites . . 80
   4. Organization of the Monophysite Church 85
   5. Persian Monophysites 90

VII. INDIAN INFLUENCE, I : THE SEA ROUTE . 96
   1. The Sea Route to India 96
   2. Alexandrian Science in India . . . 04

PAGE

VIII. INDIAN INFLUENCE, II : THE LAND ROUTE . 110
    1. Bactria 110
    2. The Road Through Marw . . . 117

IX. BUDDHISM AS A POSSIBLE MEDIUM . . . 120
    1. Rise of Buddhism 120
    2. Did Buddhism Spread West ? . . . 122
    3. Buddhist Bactria 126
    4. Ibrahim ibn Adham 130

X. THE KHALIFATE OF DAMASCUS . . . 131
    1. Arab Conquest of Syria 131
    2. The Family of Sergius 138
    3. The Camp Cities 142

XI. THE KHALIFATE OF BAGHDAD . . . 146
    1. The 'Abbasid Revolution . . . 146
    2. The Foundation of Baghdad . . . 148

XII. TRANSLATION INTO ARABIC 155
    1. The First Translators 155
    2. Hunayn ibn Ishaq 164
    3. Other Translators 170
    4. Thabit ibn Qurra 171

XIII. THE ARAB PHILOSOPHERS 176

NOTES 182

BIBLIOGRAPHY 189

INDEX 193

CHAPTER I

INTRODUCTION

THERE is a certain analogy between civilization and an infectious disease. Both pass from one community to another by contact, and whenever either breaks out, one of our first thoughts is, Where did the infection come from? In both alike there is the unanswered question, Where did it first originate?—do all outbreaks trace back to one primary source, or have there been several independent starting points?

In reading the autobiography of that distinguished orientalist Sir Denison Ross, there is a letter received from some inquirer which contains the sentence remarking what a good thing it would be if we could find out "how, and in what form, the Greek and Latin writers found their way to the ken of the Arab or Persian or Turkish student" (Sir Denison Ross, *Both Ends of the Candle*, n.d., p. 286). The author of the book makes no comment on this letter, but it may be noted that the way in which Greek literature passed to the Arabs and Persians, thence to the Turks, is not so unexplored as the letter suggests, and it may be traced with tolerable certainty, as it is hoped will appear in the following pages. No doubt it is a commonplace English convention which causes the writer to group Greek and Latin writers together : it does not appear that Latin writers ever did pass to the Arabs or other orientals, the transmission of ancient culture was concerned with Greek alone, and the Greek writers who influenced the oriental world were not the poets, historians, or orators, but exclusively the scientists who wrote on medicine, astronomy, mathematics, and philosophy, the type of scientific thought which does not always come foremost when we speak about classical literature. In the days when the Arabs inherited the culture of ancient Greece, Greek thought was chiefly interested in science, Athens was replaced by Alexandria, and Hellenism had an entirely " modern " outlook. This was an attitude with which Alexandria and its scholars were directly connected, but it was by no means confined to Alexandria. It was a logical outcome of the influence of Aristotle who before all else was a patient observer of nature, and was in fact the founder of modern science. It had its germs in older thought, no doubt, in the speculations of quite early philosophers about the origin and world and its inhabitants, animals as well as men, but it was Aristotle who introduced what may be called the scientific method.

In entering upon this inquiry it may be premised that there are at least three threads very closely interwoven. In the first place there are Greek scientific writers whose books were translated into Arabic, studied by Arab scholars, and made the subject of commentaries and summaries : in such cases the line of transmission is clear. Then there are conclusions and scientific principles assumed and developed by Arabic writers who do not say whence they were derived, but which can only be explained by reference to a Greek (Alexandrian) source. Yet again, there are questions and problems raised which the Arabs dealt with in their own way, but which never would have occurred to them unless they had been suggested by earlier Greek thinkers who had tried to solve similar difficulties, but approached their solution in a different way.

Greek scientific thought had been in the world for a long time before it reached the Arabs, and during that period it had already spread abroad in various directions. So it is not surprising that it reached the Arabs by more than one route. It came first and in the plainest line through Christian Syriac writers, scholars, and scientists. Then the Arabs applied themselves directly to the original Greek sources and learned over again all they had already learned, correcting and verifying their earlier knowledge. Then there came a second channel of transmission indirectly through India, mathematical and astronomical work, all a good deal developed by Indian scholars, but certainly developed from material obtained from Alexandria in the first place. This material had passed to India by the sea route which connected Alexandria with north-west India. Then there was also another line of passage through India which seems to have had its beginning in the Greek kingdom of Bactria, one of the Asiatic states founded by Alexander the Great, and a land route long kept open between the Greek world and Central Asia, especially with the city of Marw, and this perhaps connects with a Buddhist medium which at one time promoted intercourse between east and west, though Buddhism as a religion was withdrawing to the Far East when the Arabs reached Central Asia. Further, there were some scattered minor sources, unfortunately little known, such as the city of Harran, an obstinately pagan Greek colony planted in the middle of a Christian area, which probably made its contribution, though on a smaller scale.

The term " Arabs " must be taken in a broad sense. It is not here used strictly to denote those of Arab blood, but includes all those who were politically under Arab rule, who used the Arabic language and followed the religion of the Arabs. Some, like the Persians under the early 'Abbasids in the eighth century, were very definitely anti-Arab, but they lived under Arab rule, wrote in Arabic, and at least professed to follow the religion of Muhammad. Such being the case, they and their Arab rulers shared a common life which coloured their literature, education, and interests generally; even though Persian literature and religion diverged in its own direction, it moved from an Arabic starting-point. Neither culture nor language run on lines precisely identical with race. Conquest, superior civilization, economic needs have often caused communities to adopt new languages and new cultures. Yet there was sufficient coherence in the community gathered under the rule of the Khalif to justify its being treated as a unit, even though not all its members recognized the same khalif. The 'Umayyads in Andalus took their cue from the princes ruling in Baghdad. The schismatic Shi'ites agreed with the orthodox Sunnis that their leader on earth should be the heir of the Prophet Muhammad, though they differed as to the individual who was the lawful heir. The no less heretical Kharijites had a khalif of their own, freely elected on a democratic basis, but so elected because it was believed that this best followed Muhammad's precedent.

More important than political, racial or religious unity is the fact that those here classed as Arabs shared the same cultural history and all participated in the scientific heritage derived from the Hellenistic world. At first the city of Baghdad was the distributing centre where Greek material was brought together from different parts, Syria, Bactria, India, Persia, and other, and from Baghdad this material spread out in an Arabic form to all those social groups which were held together by the religion of Islam. Later on, when political and economic disturbances checked the cultural life of Baghdad and the empire of the khalifs began a process of devolution, or disintegration, very similar to that experienced by the empire of the Karlings in the west, the leadership passed from Baghdad to Aleppo, Damascus, Cairo, Cordova, and Samarqand. But before that happened Greek scientific literature had made itself at home amongst the Arabs and had begun a new and independent life in an Arabic atmosphere.

The Greek material received by the Arabs was not simply passed on by them to others who came after, it had a very real life and development in its Arabic surroundings. In astronomy and mathematics the work of Greek and Indian scientists was co-ordinated, and thence a very real advance was made. It may be stated that algebra and both plane and spherical trigonometry were Arab developments. The Arabs were diligent in making and recording astronomical observations, and these not only extended what they had received from the Greeks, but checked and corrected older records. The Arabs perceived the weakness in the Ptolemaic cosmology and the " new astronomy " of the thirteenth century tried to correct it, but in vain. It was not until Copernicus that the solution was found.

Not all Muslims approved of astrology. There were many who held that, as all events happen by the will of God, they could not be controlled by the stars. This was admitted, and by it came a modification of astrological theory in orthodox Islam : the stars were no longer regarded as " rulers " as in pagan astrology, but simply as " indicators " showing beforehand what God has decreed. Still some theological purists objected, and astrologers produced apologetical works to defend their science. But the Jews frankly recognized the stars as " rulers " on the ground of Genesis i, 14–16, which seems to teach that God set the lights of heaven to rule the earth, and in this were followed by the Christians.

In medicine the Arab physicians were careful observers, and their clinical records added much to what they learned from the Greeks. They invented some new instruments and in all branches, except surgery, advanced medical knowledge. Surgery was hindered by the uncleanness contracted by touching a dead body, though that impurity could be removed by the greater ablution. But there was a prevalent belief that the soul did not immediately leave the body at death, but remained in it for a period, and this caused dissection to be regarded as inhuman and cruel. From Aristotle, however, the Arabs learned the similarity in human and animal physiology, and a certain degree of progress was made in comparative anatomy. But in medicine, as in astronomy, much of their work was made obsolete by discoveries which they never knew. Harvey's discovery of the circulation of the blood, and the knowledge obtained by the use of the microscope, opened a new range of thought which threw Arab achievements into the shade. Yet for several centuries the Arab physicians were in the forefront of medical work and, as scientific progress has been continuous, their live work made its contribution not only by passing on what others had done, but by a very real development which enabled them to give to succeeding generations more than they had themselves received.

Arabic science flourished most in the atmosphere of courts. Scientists usually depended upon wealthy and powerful patrons. They appealed little to the average man, and this chiefly because scientific and especially philosophical speculation was regarded as tending towards free-thinking in religion, and so " philosophers " were classed as a species of heretics. Ultimately the philosophers themselves partially acquiesced in this judgment, and adopted the view that the inspired Qur'an was well adapted for the spiritual life of the unlettered and simple, but the illuminated saw beneath the written word and grasped an inner truth which it was not expedient to disclose to the simple.

Meanwhile, Islam generally had its own wise men, men learned in jurisprudence, tradition, and Qur'an. These were universally respected with ungrudging esteem, such as was never rendered to the scientists who were only tolerated because they were under state protection. It very much tempers our estimate of Arabic learning to remember that scientific and philosophical scholarship was confined to one privileged coterie.

CHAPTER II

HELLENISM IN ASIA

(1) HELLENIZATION OF SYRIA

HOW did Western Asia, what is now often called the Near East, come under Greek influence? The starting-point was Alexander's conquest of Persia in 331 B.C. The great oriental kingdom of Persia, which stretched from the Indus to the Mediterranean, went to pieces before the attack of this prince who was ruler of one of the comparatively petty states of Greece. It is one of the many instances in history showing that vast numbers count for little when faced by a small but thoroughly efficient force. The Greeks followed up this victory by an invasion of Persia which gradually brought the whole country under their control and at length penetrated as far as the Punjab, which was claimed as a Persian province. This political conquest did not result in the whole conquered territory becoming Greek, it remained Persian under Greek rule, Alexander planting colonies in the nature of Greek garrisons scattered here and there in the conquered land.

Alexander died, yet a young man, in June 323, leaving only an infant son as his heir. Immediately his generals began quarrelling over the heritage, and these civil wars lasted until 312 when the leading competitors consented to divide the spoils, and in this division Seleucus obtained the Asiatic share, practically the whole of the old kingdom of Persia. But Seleucus was jealous of another general, Ptolemy, who had obtained Egypt, and was much more concerned with his rivalry with this Egyptian monarch than with the internal affairs of Persia. About 300 B.C. he built a new capital Antioch in Western Syria and left the main part of his Asiatic territory in the hands of a deputy. Profiting by this a new independent kingdom of Parthia was formed by Arsaces in 248, much smaller than the old Persian kingdom but still a great power, and before long this began encroaching on the Seleucid heritage. Gradually it crept nearer and nearer to the Mediterranean until in 150 B.C. it absorbed Mesopotamia and the Seleucid state was reduced to little more than Syria. Thus Greek control was permanent only in the area bordering on the Middle Sea.

How far did this country under Greek rule become Greek? This is best illustrated by parallel conditions in Egypt. In the dry clear air of Egypt documents of the Ptolemaic period have been preserved in large numbers, and from these we can learn a good deal about the Hellenization of the country, whilst in the humid climate of Syria such documentary material is comparatively rare. From Egypt we learn that all official business was conducted in Greek, and it was necessary for anyone aspiring to a post in the civil service to know Greek. Manuals still exist to help aspirants to acquire a knowledge of the Greek language and material survives to show how far they succeeded in doing so. Apparently the Egyptians found Greek a very difficult language and in most cases their mastery of it was very defective. It is quite clear that it never really became the language of the country. Egyptian was used in the home and in the markets, only those who wished for government employment tried to get a command of Greek. Even in Greek colonies like Alexandria and Coptos, where Greek was the language of the citizens, there was a large class, mostly occupying its own quarter of the city, which used only the native speech. In Greek cities the citizens formed only a privileged ruling class, often a minority. Outsiders (metics) who settled in the city and persons of the native subject population, as well as slaves, had no rights as citizens. Thus the Greek language, and with it Greek culture, customs, and religion were confined to the ruling class and had very little influence on the people of the villages, the tillers of the soil, and the subject community generally. Then again there often was intermarriage, and the home generally used the vernacular and inclined to sink back into native ways. This seems to have applied equally to Syria. The Greek language was used by the ruling class in the greater towns, it was used by officials throughout the country, but it produced only a Greek surface beneath which the native population remained, not unaffected by Greek influence but affected only slightly by it.

The usual language of Syria and Mesopotamia was Aramaic, a language akin to, but by no means identical with, Hebrew. The name 'Aram signifies highland, and Aramaic generally was the language of the higher country in the north and in the hinterland, whilst Hebrew was used in the lowlands and came closer to the Phœnician language used on the littoral. But Aramaic had a good many dialects, as it spread over a very wide area. In later times one important dialect, or group of dialects, developed amongst the Christian population of Syria and Mesopotamia, with its centre at Edessa, came to be known as Syriac, and this Syriac-Aramaic [1] was the chief medium by which Greek culture was passed on to the peoples of the Near East. In oriental lands communities most often rest on a religious basis : nations are only temporary groups formed for political purposes, religions form social groups which share a common cultural life, economic structure, literature and art. As a rule the barrier between men of different religions is more definitely marked than that between members of different political states.

In the middle of the second century B.C., when the Parthians conquered Mesopotamia, the Seleucid state was decadent, worn out by a long and futile struggle to get control of Egypt. The Parthians did not follow up their conquest, because by that time they were being attacked in their eastern provinces by Mongolian tribes, and had no military resources to spare for the west. But there was a third power close at hand which was able to take advantage of the weakness of Syria, Armenia under an ambitious monarch Tigranes, and he conquered Syria in 83 B.C. But by this time a new power had appeared on the shores of the Mediterranean, the Roman Republic, not a conquering power like that of Alexander, but a rather narrow-minded democracy whose chief aims were to carry on trade successfully and make sure of safety at home. For safety the Romans gradually carried out the conquest of Italy, then they tried to exercise a kind of protectorate over all the other countries around the Mediterranean, and to check any one which tended to interfere with its safety or commerce. Conquest and expansion were forced on Rome by circumstances, and were undertaken by Rome only when foreign rivals threatened its security or its commerce by commercial rivalry like Carthage or by piracy on the seas over which Roman commerce passed, as was the case with Pontus.

Italy, a long narrow peninsula with a protracted coast line

necessarily depended on sea power for its own security as well as for international trade, though that received only a tardy and grudging recognition in Rome. Gradually it was perceived that the freedom and prosperity of Italy, which included that of Rome, depended on control of the Middle Sea, and necessitated a check on the formation of any great power along its shores which could intercept sea communications. An attempt at founding such a power was made in 168 B.C., when the Seleucid Antiochus Epiphanes made an attempt to conquer Egypt. He was camped before the walls of Alexandria when an envoy arrived from Rome warning him to retire, and that he reluctantly did. Rome was already a formidable power, and the Seleucid considered it wiser not to challenge it. Next, Mithridates VI of Pontus formed imperial ambitions. He occupied Asia Minor, massacred a number of Roman citizens, and then invaded Greece, whilst Pontic pirates ranged over the eastern Mediterranean. The Romans had no wish to interfere in eastern politics, but this forced them to do so, and the Mithridatic War followed, which the Romans under Pompey brought to a successful conclusion in 83 B.C. These events forced Rome into the tangled political strife of what we now call the Levant, and in 81 B.C. they were still further drawn in when Alexander II of Egypt died and left his kingdom by will to the Roman people.

Syria had by then long ceased to be a danger. Parthian control had passed away from Mesopotamia and Syria, as the Parthians had to deal with threatening pressure on their own eastern borders. Under the degenerate Seleucids Syria was near a state of anarchy. The real masters of the country were the Arab tribes, many of them roaming the country as brigands, others settling down in lands they conquered and forming native states.

Pompey had just completed the Mithridatic War when the last Seleucid monarch Antiochus Asiaticus came to the throne, and thought it expedient to obtain formal recognition from Rome. To his request Pompey replied that Rome would not recognize any monarch who could not keep his country in order, and by now it was obvious that the Seleucids could not do this. So in 65 B.C. Syria was annexed and made a Roman province under a legatus whose first duty was to defend the frontier against the Parthians, Pompey determining that the

River Euphrates should be recognized as the frontier. But the Arab states formed along the eastern borders of Syria were left alone, and so the larger state known as Nabataea, though in 63 Pompey led an expedition against the Nabataean capital Petra. Thus Syria passed out of Greek Seleucid control and became part of the Roman Empire. Politically it was a change, but culturally there was no change, the influence of Rome was as definitely Greek as that of the Seleucids had been. The cultural life of Syria and Mesopotamia went on unaffected by the political change and from that time forward it was the Romans who brought Greek influence to bear on the Near East.

(2) THE FRONTIER PROVINCES

When Syria became a Roman province it was secured against the immediate menace of its two oriental neighbours, Parthia and Armenia. Roman arms protected the border and sometimes crossed victoriously into enemy territory. But with this began a long series of wars lasting for some seven centuries, in which the frontier frequently shifted according to the fortunes of war. There was a debatable territory between the Tigris river and the Libanus mountains, which was sometimes Græco-Roman, sometimes Parthian or Persian, and these political vicissitudes had their effect on the cultural life of the area involved.

The Emperor Augustus recognized the Euphrates frontier and allowed the Arab states to remain without interference, and so matters continued until the accession of Trajan, though the trade route through Mesopotamia was practically closed because the Parthians were unable to control the tribesmen along the border. Trajan decided to carry Roman authority farther east and to bring the disordered border lands into a more satisfactory condition, and to effect this in A.D. 115 conquered Mesopotamia and made it a Roman province. The following year he invaded Parthia, advanced to the Tigris, occupied Adiabene in northern Mesopotamia and made it a province under the name of Assyria, took Seleucia the chief Greek colony on the Tigris and the Parthian capital Ctesiphon close by, and went on as far as the mouth of the Tigris, but was called back by the news that Mesopotamia in his rear had revolted. That revolt he put down, burning Seleucia and

Edessa, but died on 8th August, 117.   His policy was reversed by his successor Hadrian, who gave back Mesopotamia and Assyria and resumed the Euphrates frontier, whilst Armenia which had also been annexed ceased to be a Roman province but remained a vassal state.

As soon as Antonius Pius died in 161, the Parthians invaded Armenia and placed an Arsacid prince on its throne, then they invaded Syria and defeated the Roman army there.   This forced the Romans to act, and Verus, who was co-emperor with Marcus Aurelius, went east to command the army in person in 162.   Though the Parthians stubbornly defended the Euphrates, the Romans at length broke through, advanced into Mesopotamia, besieged Edessa and Dausara, and approached Nisibis the frontier fortress, then took and destroyed the Parthian capital Ctesiphon.   But the victorious army brought back the plague with it, and from that many perished.   At the end of the campaign Rome secured the western half of Mesopotamia, the prince of Edessa became a Roman vassal, and the town of Harran was made free under Roman protection.

In 194 Septimus Severus led a Roman army into Mesopotamia, the whole of which he made a Roman province, as it had been under Trajan : Nisibis was made the capital of this province, and Edessa was allowed to continue as a vassal state.   But in 198 the Parthians resumed hostilities and advanced into Mesopotamia, sweeping all before them until they reached Nisibis to which they laid siege.   The Emperor Severus had started his return journey, but this recalled him : he rescued Nisibis and proceeded into Parthia where he took Seleucia and Ctesiphon from which the Parthian king escaped with a few horsemen, leaving the royal treasure for the Romans.

This defeat told severely on the Parthians and brought about a revolt in 211, which ended in dethroning the Arsacid dynasty and restoring a kingdom of Persia under the rule of the Sasanid family which claimed descent from the ancient Achaemenid kings.   In the east political movements most often have a religious bearing, and this Sasanid revolution was associated with a revival and reform of the Mazdean religion.[2] Anciently the Persian kings had belonged to a priestly caste, and were regarded as endowed with a divine spirit, but the

[2] See note on p. 182.

Parthian monarchs were not of this privileged order. In the course of the first century of the Christian era, it would appear, some of the Parthian rulers had tried to lead a religious reformation, but their caste inferiority had hindered their efforts. Since then religious observances had been relaxed : the sacred fire had been allowed to go out (Moses of Chorene, *Hist. Armen.*, ii, 94), the fire had been defiled by the fact that the bodies of the dead had been burned contrary to Mazdean religious law (Herodian, iv, 30), and the priestly caste of Magi had fallen into disrepute (Agathias, ii, 26). No doubt the impression was that a restoration of the old semi-divine monarchy would bring about a revival of national greatness.

The Sasanid placed on the restored Persian throne was Ardashir, and one of his first acts was to hold a general council which dealt with internal divisions which had caused the Mazdean religion to separate into several sects, and so to form an established state church. On the one side the religious revival which had been gathering force for some years was completed, and on the other side the king undertook to restore the military prestige of the country which had suffered so great an eclipse under the later Arsacids.

From 224 to 241, Ardashir was occupied in putting down the adherents of the displaced Arsacid dynasty, but in the course of-that period, in 230, he sent a challenge to Rome demanding of the Emperor Severus that all the territory which had ever been in the hands of Persia should be restored to him, Syria, Asia Minor, and Egypt, and at the same time made preparations for the invasion of Syria. This, of course, was a declaration of war. But Ardashir was unable to proceed further immediately, as he had not yet effectually reduced the pro-Arsacid party, then in 241 he died, leaving the kingdom and the war to his son Shapur (241–272). The outbreak of war was hastened by events in Armenia, where king Khusraw, a member of the Arsacid family who had been placed on the Armenian throne by the Romans, was assassinated by emissaries of Shapur. The Armenian nobles, however, refused to support Shapur, and declared in favour of Khusraw's younger son, Tiridates, who was a ward of Rome. Then Shapur occupied Armenia and Tiridates fled to the Romans. From Armenia the Persians overran Mesopotamia, Cappadocia, and Syria, where they took and plundered Antioch, but were held up before Edessa. Then the Emperor Gordian advanced against the Persians, defeated them, and drove them back. This restored Roman rule as far as the Tigris, and Gordian proceeding farther threatened the Persian capital Ctesiphon. But Gordian was murdered in 244, and his successor Philip made a peace which gave Armenia to Persia, Mesopotamia to Rome.

War broke out again in 258. At that time the Roman Empire was under the Emperor Valerian and his son Gallienus. Shapur repeated his former tactics of 241, and Valerian prepared to invade Persia. He occupied Cappadocia, the Persians retiring before him, but the plague played havoc with the Roman army, and delayed it too long before entering Mesopotamia. Near Edessa some time in 259–260, the exact date cannot be determined, he met the Persians and was totally defeated, both he and his army taken prisoners. He remained a captive in the hands of the Persians until his death in 267. The Persians then swept through Syria and captured and plundered the city of Antioch. The only check they received was from a self-appointed commander named Callistus, who sailed with ships from the harbours of Cilicia and went to the relief of Pompeiopolis, which the Persians were besieging, killed several thousand men, and took possession of Shapur's harem. This caused the Persian king to turn back and hasten home, paying to Edessa all the plunder he had taken from the Romans for permission to pass unmolested through their territory. During this retirement, the Persians were harassed and suffered heavy losses at the hands of Odaenathus, King of Palmyra. After this, two leading Romans, the Callistus who had relieved Pomeiopolis and Macrianus the army paymaster, renounced allegiance to Valerian's son Gallienus, and proclaimed Macrianus' two sons, Fulvius Macrianus and Fulvius Quietus, as joint emperors (261). These two were recognized in Egypt and the east, with the exception of Palmyra, which remained loyal to Gallienus. But Fulvius Macrianus went west, and fell in battle with another pretender, whilst Fulvius Quietus was betrayed by Callistus and put to death by Odaenathus. Thus unexpectedly Palmyra and its ruler Odaenathus became dominant factors in the politics of the Near East.

(3) Foundation of Jundi-Shapur

Many of the prisoners taken by the Persians from Valerian were sent to work constructing the Great Weir or *Shadurwan* across the Dujayl river below Tustar, of which portions still remain. Those prisoners who were men of education or technical skill were treated generously, for Shapur recognized the superiority of the Romans in these respects, and hoped to employ such prisoners as engineers, architects, physicians, land-surveyors, and the like. He settled these educated captives in three cities where they were allowed to live according to their own laws, using their own language and following their own religion. One of these cities was close by Susa, the Shusan of the Old Testament (Dan. viii, 2 ; Nehem. i, 1 ; Esther i, 2), which was one of the royal residences and served as the king's winter palace. The prisoners' camp-city near Susa was named *Beh-an-Andew-i-Shapur* " Shapur's Better than Antioch " (at-Tabari, *Ann.* ii, 861, 6), or *Jundi-Shapur* " Shapur's camp ", but the Syrians called it *Beit Lapat* " the House of Defeat ". " Eight leagues north-west of Tustar, on the road to Dizful, lie the ruins now called Shahabad, which mark the site of Junday Sabur or Jundi-Shapur. Under the Sassanians Junday Sabur had been the capital of Khuzistan " (Le Strange, *The Lands of the Eastern Caliphate*, 236). As Susa was the winter residence of the Persian kings, we read that " all the Sasanid kings which we have mentioned up to Hormuz the son of Narsai dwelt at Jundi-Shapur in Khuzistan " (Mas'udi, *Muruj*, ii, 175).

As the captives were free to follow their own religion they enjoyed greater religious freedom under Persian rule than they were officially permitted at that time in the Roman Empire, for those who were Christians were allowed to build and maintain churches, whilst within the Roman jurisdiction Christianity was still liable to persecution. At Yaranishahr, which was one of the camp cities assigned the captives, they had two churches, in one the liturgy was celebrated in Greek, in the other the Syriac language was used (*Chron. de Séert*, ed. Scher, in P.O., iv, 220–1).

There is a tradition that the Bishop of Antioch, Demetrianus, was one of the captives, and was asked by his fellow-prisoners to act as their bishop, retaining the title of Bishop of Antioch, but this he refused : then the Catholicus Papa made him bishop of Jundi-Shapur and gave him the first place at the consecration of a catholicus, which was the title given to the Bishop of Seleucia as primate of the Persian Church. But this tradition is based on Mare's *Liber Turris* (p. 7), a late work and one which " fourmille d'invraisemblances et d'anachronismes " (Labourt, *Le Christianisme dans l'empire Perse*, 20, note 1). The writer seems to have supposed that the Bishop of Antioch (not yet called Patriarch) was one of the dignitaries of the imperial court, which could hardly have been the case under Valerian, and that the church at that early date already was fully organized with patriarchs, archbishops, and metropolitans, all a post-Nicene development.

(4) DIOCLETIAN AND CONSTANTINE

After their defeat in 260 the Romans, beset by many enemies, were for some time unable to take steps to recover their authority in Asia, and for a while Palmyra enjoyed special prestige. The city was a Roman ally, but not under Roman protection. Its subject territory reached to the Euphrates and included the important crossing at Sura. Since the disordered period at the end of Seleucid rule it had become the chief mart on the trade route between Mesopotamia and Syria, and so very wealthy. It had adopted Græco-Roman art and architecture, but remained very much an oriental power. Greek inscriptions at Palmyra are rare, but the Aramaic inscriptions giving public decrees often had a Greek translation attached. It retained its native deities and used a calendar which reckoned by what are known as the Assyrian months.

After 260, Odaenathus assumed the title of king, and occupied the position of an independent viceroy under the more or less nominal suzerainty of Rome. In 264, he crossed the Euphrates, relieved Edessa, and recaptured Nisibis and Harran (Carrhae) from the Persians, then marched into Persia and attacked Ctesiphon. For the time he was independent and important, only nominally under Roman authority. But in 266–7 he was murdered, not as was suggested, at the instigation of a jealous Roman government, but by a treacherous nephew influenced by a private quarrel.

At Odaenathus' death the government of Palmyra was assumed by his widow Zenobia, who thereby claimed to rule over Egypt and Asia, though in fact her power was limited to Syria and Arabia. She made an attempt to enforce her authority in Egypt, and in face of a sturdy opposition, conquered the country, whilst in Asia she extended her authority to Chalcedon in front of Constantinople. Whatever profession of loyalty to Rome there might be, Palmyra had become a rival and hostile power. In 270 Aurelian (270–275), an energetic and capable prince, dislodged the Palmyrenes from Egypt and went to Syria, thence advancing eastwards towards Palmyra. The Palmyrenes were defeated with heavy loss on the banks of the Orontes near Antioch, and again when they made a stand at Hemesa, then the Romans marched across the desert to Palmyra itself. At this Zenobia lost her nerve, and fled to seek refuge with the Persians, but was overtaken and brought back a prisoner ; whereupon Palmyra surrendered (272). Next year it revolted, but Aurelian turned on it with unexpected rapidity, took the city, and destroyed it. Thus Roman rule was restored in Syria.

Meanwhile Shapur I of Persia had died (271), and was succeeded by his son Hormuz I, who had only a brief reign of one year and ten days, and was followed by Bahram I (272–273). In his days appeared the heretic Mani, founder of the Manichaean sect, and the king had him executed as an offender against the Mazdaean religion. Either he was crucified at Jundi-Shapur, or his body was flayed after death and the stuffed skin exposed on the gate of that city, in any case there was a connection with Jundi-Shapur (at-Tabari, *Ann.* ii, 90 ; Scher, *Chron. de Séert*, P.O. iv, 228). In 273 Bahram sent help to Zenobia, but not sufficient to save her, and by this provoked the enmity of Rome, but he was not prepared for war, and sent an embassy to conciliate. The Emperor Aurelian (270–275), however, was determined to enter on a war with Persia to wipe out the disgrace of Valerian, and this was popular with the Roman people, but before action was taken Aurelian was murdered (275).

Bahram I had been followed on the Persian throne by two other kings of the same name, Bahram II (273–276) and Bahram III (276–293). These were succeeded by Narsai (293–302).

After various vicissitudes in the Roman Empire, Diocletian ascended the imperial throne in 284. In the course of his reign, in 296, Narsai declared war against Rome under pretext of enforcing his claim to Mesopotamia and Armenia. Diocletian sent his colleague Galerius, and this time the Romans won a decisive victory, and in 298 a satisfactory peace was concluded, by which the River Aboras was recognized as the boundary between the two states, five provinces beyond the Tigris were ceded to Rome, and the pro-Roman prince Tiridates was confirmed as King of Armenia.

Constantine, who succeeded Diocletian in 306, reigned until 327, and Shapur II (309–370), who had become king of Persia, observing the many difficulties gathering around Rome, in 359 invaded Mesopotamia and besieged Amida, which he took after a long siege. It was inevitable for Rome to interfere again, more especially because repeated efforts had been made to capture the great frontier fortress of Nisibis. In 362, the Emperor Julian, at the head of a large army, invaded Persia, but this enterprise turned out ill : he himself was slain, his army was defeated, and it was only with great difficulty that his successor Jovian rescued its remnants. After this disaster the Romans were compelled to purchase peace on very unfavourable terms, and the five provinces ceded to Rome in 298 had to be restored.

It was under Hormuz that Jundi-Shapur had ceased to be a royal residence, and gradually became a heap of ruins. Shapur II, his successor who repelled Julian's invasion, took many prisoners in his war with the Romans, and " left the land of the Romans bringing away with him captives whom he settled in the lands of 'Iraq, al-Ahwaz, Persia, and the cities built by his father. He himself built three cities, and called them after his own name. One of these was in the land of Maisan, and was called Sod Sabur, now it is called Der Mahraq. The second, in Persia, is still called Sabur. He rebuilt Jundi-Shapur which had fallen into ruins, and called it Anti-Shapur (Antiochia Saporis) . . . the third town is on the banks of the Tigris, and he called it Marw Haber, now called Akabora " (Scher, *Hist. Nestorienne (Chron. de Séert)* in P.O. iv, 221). Later writers such as Abu l-Farag, often refer to Shapur II as the founder of Jundi-Shapur, but the more correct view seems to be that the city was founded by Shapur I, that it fell into decay when the court left the vicinity in the days of Hormuz II, and that it was rebuilt by Shapur II.

So far the diffusion of Hellenism was the work of the Seleucids, then of the Romans. Now a new factor appears. In the fourth century the eastward spread of Hellenism became the deliberate task of the Christian Church, which at that time identified itself with the Roman Empire. From this point the political history of Rome may be laid aside and attention concentrated on the outspread of Christianity.

CHAPTER III

THE LEGACY OF GREECE

(1) ALEXANDRIAN SCIENCE

POLITICAL events had brought western Asia a good deal under Greek influence. There had been some centuries domination of the Seleucid kings of Syria and, though the later rulers of that dynasty were inefficient and weak, the earlier ones had been otherwise. Public business had been carried on in Greek, and all who aspired to share in the administration had to learn and use Greek. No doubt this Hellenization was superficial, we know that it was so, but it left its impress. Then came Roman rule, which brought no new culture but rather reinforced the already existing Greek influence. Finally came the Christian Church, which was more definitely Greek in its influence than either the Seleucid kings or the Roman State, and after the time of Constantine the Roman government and the Christian Church worked hand in hand.

But the Greek culture which was thus introduced, was not that of Athens. Its focus was Alexandria in Egypt. It was not Hellenic, but Hellenistic. No doubt the culture of Alexandria evolved naturally and indeed inevitably from that of the older Greece, but it took a rather different direction. Philosophy as it was down to the age of Plato began to specialize in natural science under the guidance of Aristotle, and ultimately concentrated itself in medicine, astronomy, and mathematics. All these were treated as phases of natural science, and philosophy dealt with the underlying realities of which these specialized sciences were regarded as aspects. Its aim was to get the key to the natural order which, it was believed formed one great harmonious whole, and the means to be employed in the inquiry was outlined by the strict use of logic. This, of course, meant that the methods used in science held good in theology also, and this assumption caused the Church to be a missionary of Greek intellectual culture as well as of the Christian religion.

The City of Alexandria had been founded by Alexander the Great in 323 B.C. Its site was already occupied by the Egyptian town of Rakote (ρακοτε) and this continued to be the name of the city in the Egyptian vernacular Coptic. When Alexander's empire was divided amongst his generals, Egypt was secured by Ptolemy Soter, and remained in the hands of the Ptolemaic dynasty until it was taken over by the Romans. Ptolemy Soter made Alexandria his capital, and did much to render it the focus of Greek culture and scholarship. He founded there the Museum which before long became a kind of Hellenic university, a rival of the older schools of Athens. Apparently there had been a kind of congregation of sages in the temple of Heliopolis before this, and these sages removed to the new foundation which thus became an heir of the wisdom of the Egyptians. But the Egyptian element seems to have been absorbed in the Greek atmosphere, so that Alexandria was the heir of Athens rather than of Heliopolis. Still, the Greek world of Alexandria lost the exclusiveness which had marked Athenian thought. It took a cosmopolitan character and showed a marked leaning towards oriental thought. In spite of its professed exclusiveness, earlier Greek culture had not been quite free from oriental influences, and much that appears in Greek life and thought can be traced back to Egypt and Babylon. Again, it must be noted that although Alexandria became so prominent in the development of later Greek thought, such development was not confined to it ; it was not local, nor even national, but cosmopolitan. The Egyptians themselves never reckoned Alexandria as a part of Egypt ; to them it always was a Greek colony, the headquarters of the alien race which garrisoned and ruled Egypt.

The Museum was founded by Ptolemy Soter who attached a library, but it was the generosity of his successor Ptolemy Philadelphus (285–247 B.C.) which enriched this until it became the greatest library of the ancient world, and this by itself went far to make Alexandria a gathering place of the learned.

The new cosmopolitan Greek life which developed after the days of Alexander had many sides. It produced its own class of literature, and evolved a scientific literary criticism. It carried forward philosophy, often on rather new lines. It produced new research in medicine, astronomy, mathematics, and other branches of science. All these were inter-related, for all show a kindred spirit, and all evolved naturally from the culture of the older Greece. But, as a matter of convenience, it will be well for us to concentrate our attention on three leading branches, philosophy, medicine, astronomy, and mathematics, these two last regarded as one because closely allied and developed at the hands of the same persons.

(2) PHILOSOPHY

Aristotle the philosopher had been Alexander's tutor, but his life was more connected with Athens than with Alexandria. Yet his influence permeated Greek thought, and was mainly responsible for directing it towards natural science and mathematics, though this scientific tendency had a precedent in earlier philosophy.

The latest type of Greek philosophy, and one which had very great influence on Greek thought when it came into contact with the Arabs, was that known as neo-Platonism. This school of philosophy was fond of tracing its beginnings back to the semi-legendary Pythagoras (580–500 B.C. ?), a native of Samos or of Tyre who, if not the pupil of Thales, at least visited him and was influenced by him. Thales is said to have studied mathematics and physical science in Egypt, and Pythagoras is described as following in his footsteps and going to Egypt and receiving instruction there from the priests. Amongst other things he learned from these priests the doctrine of transmigration (cf. Herdt. ii, 123). On returning home he found that Samos was under the tyrant Polycrates, and thereupon migrated to Magna Graecia, ultimately settling at Croton. There he established a school in the form of a confraternity, following Egyptian precedent. This fraternity possessed all its goods in common, and kept all its teaching secret from the outside world, which caused it to be regarded with suspicion, as a secret society with potential subversive political tendency. So the fraternity experienced rough treatment and Pythagoras escaped to Tarentum, then to Metapontum. The community was broken up, but continued as a philosophical group for some two centuries, though no longer preserving secrecy about its tenets. The rule of secrecy was first broken by Philolaus (circ. 400 B.C.), in fact such secrecy was altogether alien to Greek thought. After the fourth century B.C., when Philolaus disclosed its esoteric doctrine, the Pythagorean school declined in prominence. Pythagorean schools or clubs in Magna

Graecia had assumed a political character, strongly anti-democratic in their tone, and at some period in the course of the fourth century there was a rising against them during which the cities of Magna Graecia became a scene of murder, armed rebellion, and disorder of every kind (Polybius, ii, 39 ; Strabo, viii, 7, 1 ; Justin, xx, 4). Plato shows tendencies towards Orphic and Pythagorean ideas, especially in the later treatises. The Old Academy was more Pythagorean than Plato, but the New Academy turned in a different direction. Whether the doctrine of the immortality of the soul came from Egypt through a Pythagorean medium is not clear, but most of the Greeks who accepted that doctrine were in touch with Pythagoreanism.

About 100 B.C. there was a revival of Pythagoreanism and a number of pseudonymous treatises appeared purporting to describe Pythagoras' teachings, including a set of poetical maxims which were called " the Golden Verses of Pythagoras ". It does not seem that the Pythagorean school ever took root in Rome. In this maturer Pythagorean teaching the soul was regarded as consisting of three parts, *nous*, *thumos*, and *phrenes*, only the first of these immortal. All nature was regarded as being alive, animated by heat, and the sun and stars as centres of heat were esteemed to be gods. The movements of the heavenly bodies are harmoniously adjusted by number, an idea of Egyptian origin, and so certain numbers have a sacred character, e.g. 10 which represents the sum of a pyramid of four stages, $4 - 3 - 2 - 1 = 10$. This consideration of numbers appears again in Philo and later philosophers. All these ideas recur again in the later neo-Platonic philosophers, whose influence was felt by the Arabs. From the beginning Pythagorean teaching was much concerned with mathematics, its geometry chiefly interested in measuring areas. The Athenian Sophists turned to the geometry of the circle which the Pythagoreans had neglected. This revived Pythagoreanism exercised great influence in later Athens, and apparently in Alexandria as well. Neo-Platonists knew Pythagorean teaching in this later form. Both Porphyry and Iamblichus, leading neo-Platonists, wrote lives of Pythagoras. In itself neo-Platonism was a perfectly natural and logical development of Greek thought, not an oriental intruder. It was eclectic, but so were most of the later philosophies, and combined the systems of

Plato, Aristotle, and the Ṣtoics under the ægis of Pythagoras.
It received its clear definition in the teaching of Plotinus and his
disciples.

The Neo-Pythagorean philosopher *Numenius* of Apamea (circ.
160–180 B.C.), whose teaching is known by citations in Eusebius
(*Praep. Evang.*, xi, 10 ; xviii, 22 ; xv, 17), and a few other
references (e.g. Porphyry in Stob., *Eccl.* i, 836) must be
regarded as a precursor of neo-Platonism. He was the first Greek
philosopher to show any sympathy with Hebrew religion,
describing Plato as Moses speaking in Attic (Clement Alex.,
Strom. i, 342 ; Eusebius, *Praep. Evang.* xi, 10). He shows very
plainly a tendency to religious syncretism such as is strongly
marked in the neo-Platonists, but is not confined to them, indeed
it seems to have been widely prevalent in the second century
and after.

The neo-Platonic school had its parent in *Ammonius Saccas* or
Saccophorus, so named because he had been a carrier in his
youth. Very little is known of his life. The chief source of information is Porphyry cited by Eusebius (*Hist. Eccl.* 6, 19, 7),
who states that he was a native of Alexandria and a Christian
educated by his parents in the faith, but when he began to study
philosophy he changed his opinions and became a pagan,
though this last statement Eusebius denies (ib. 6, 19, 9). It has
been suggested that Eusebius confuses him with another
Ammonius, his contemporary and also an Alexandrian who was
the editor of a Diatessaron giving the gospel according to
St. Matthew with parallel passages from the other gospels, the
basis of what afterwards were known as the Ammonian sections.
Hieronymus (*de vir. illust.* 55) says, " de consonantia Moysi
et Iesu opus elegans et evangelicos canones excogitabit ".
Apparently there were two contemporary persons, both of
Alexandria and both called Ammonius. According to Longinus
and Porphyry our Ammonius refrained from writing any books,
following the precedent of Pythagoras, but the other Ammonius
was the author of several works. Amongst the pupils of
Ammonius were Origen, Plotinus, Herennius, Longinus the
critic, Heracles, Olympius, and Antonius, but these may not all
have been pupils of the same Ammonius. Porphyry says that
his teaching was kept secret, also a Pythagorean idea, that he
bound his pupils by oath not to disclose it, but that vow was
broken first by Herennius, then by Origen. There were two

Origens, one the well-known Christian writer, the other a pagan philosopher, both Alexandrians and contemporary, but the Christian Origen and Heracles may have been the pupils of the other Ammonius who composed the Diatessaron. As to Ammonius' teaching, Hierocles (apud Photius) says that he endeavoured to reconcile Plato and Aristotle, but that was the aim of all the later Alexandrians. Nemesius, a bishop and neo-Platonist of the later fourth century, gives two citations, one from both Numenius and Ammonius, the other from Ammonius alone, both about the nature of the soul and its relation to the body. If it be true that Ammonius did not leave any writings, these references can only represent traditions about his teaching. The association with Numenius is significant.

Plotinus was an Egyptian, a native of Lycopolis or Siut, now known as Assiout, where he was born about A.D. 200 (Eunatius, *Vit. Soph.* p. 6 ; Suidas, *sub voc.*, puts his birth at Nicopolis). He attended the school of Alexandria, but was dissatisfied with the teaching he heard there, until a friend took him to hear Ammonius Saccas. On hearing his lecture, Plotinus decided that he had found the right teacher. He was then in his twenty-eighth year, and remained with Ammonius eleven years. Undoubtedly the meeting with Ammonius was a turning-point in Plotinus' life and gave the clue to his doctrine. But Ammonius wrote no books, nor did he make any effort to publish his teaching, preferring to instruct in private and under a pledge of secrecy. One result of Ammonius' teaching was to make Plotinus anxious to obtain more accurate information about the beliefs of the Indians and Persians. Reverence for, and interest in, oriental thought was characteristic of the Alexandrian school and this was inherited by the neo-Platonists. In order to gratify this desire Plotinus joined the Emperor Gordian's expedition to Persia in 242, an expedition which turned out ill and resulted in the emperor's death, and Plotinus had difficulty in reaching Antioch in safety. He then went to Rome, being at the time forty years of age, and there lectured for ten years and had many hearers, some of them senators and other leading citizens. But for long he followed Ammonius' example and taught privately, writing and publishing nothing. Then in 254, he began to write, and in 263 Porphyry became one of his hearers, introduced by Amelius who had been his hearer for twenty-four years, and remained with him six years. Plotinus had written twenty-one books of his *Enneads* when Porphyry met him, during the six years they were together he wrote twenty-four more, which Porphyry considered his best work, and in the brief remainder of his life he wrote nine more. He died in 269, having completed his 69th year. His death took place during a visitation of plague, but was not due to the pestilence. Apparently he became ill because he was deprived of the ministrations of his personal attendants who had been carried off by the plague. Finding himself ill, he retired to Campania to a house bequeathed to him by the Arab physician Zethus, who had been one of his pupils, and there finished his life in peace.

Later neo-Platonists often associated themselves with the revival of paganism then in progress, as did his pupil Amelius, but Plotinus himself stood aloof. The *Enneads* have come down to us rearranged and revised by his pupil Porphyry who, however, outlines another arrangement disposing the books in chronological order, and by that arrangement the development of Plotinus' thought is made clearer.

Though Plotinus was educated at Alexandria, his teaching was developed and delivered in Rome. At one time neo-Platonism was regarded as essentially Alexandrian, but this is an overstatement, if not altogether untrue, though the system contains elements which appear also in the Alexandrian Jew Philo, in the Gnostics who seem to have been of Egyptian origin, and in the Alexandrian Christians Clement and Origen. It was indeed eclectic, though claiming to be Platonism. It had a religious syncretism akin to that which appears in Plutarch and Maximus of Tyre, and which seems to have been very widely prevalent at the time.

In Plotinus' teaching the Monad is presented as the Supreme God, the ultimate source of all good and order. God is immanent, but is also transcendent. Between God and the world is the World Soul, the creator whose work is not altogether good and orderly, whilst the phenomenal world itself is unsubstantial and unstable. It is very much like the Gnostic attitude towards the problem of evil : the Creator whose work is obviously imperfect, is a subordinate, not the Supreme God, and therefore not perfect. Knowledge may be obtained by sense-perception, by inference from sense-perception, but the highest and best knowledge is that received directly by inspiration.

c

Neo-Platonism, substantially the doctrine of Plotinus' *Enneads*, though developed by his successors, exercised a powerful influence over the Graeco-Roman world for several centuries. Books IV–VI of the *Enneads*, in an abridged Syriac translation, circulated amongst Syriac-speaking Christians, especially the Monophysites, as the " Theology of Aristotle " and were accepted as genuinely Aristotelian by the earlier scholars of Baghdad, before the time of al-Kindi, and were still so accepted by many for long afterwards. It is easy to see how such material contributed to a pantheistic and mystical tone of thought such as is apparent in Muslim philosophy.

Porphyry (b. 233, died after 301) was a Syrian, his original name Malchus meaning " king " or " royal ", which he changed at the advice of his teachers to Basileus, then to Porphyry. He studied at Athens under Longinus, Ammonius' disciple, then at Rome in 263 under Plotinus. After a visit to Sicily he returned to Rome and gave expository lectures on the philosophy of Plotinus. He married Marcella, a friend's widow, simply for the sake of educating her children. At the time there were many sects which produced spurious apocalyptic works which they attributed to various distinguished authorities of ancient times, and with some of these Porphyry was led into controversy, especially against a book published under the name of " Zosimus " and purporting to give an account of the religious tenets of the Persians. This work he showed to be a recent forgery, and in doing so applied sound principles of criticism. The inquiry led him into controversy with the Christians, and for several centuries his writings were viewed by the Christians as the most serious attack made upon their faith. Only fragments of his work in this direction are preserved by Christian apologetical writers, but it is clear that his method of treatment was by way of historical criticism as already developed in the school of Alexandria. In one treatise, *De antro nympharum*, he applied the method of allegorical interpretation to the story of Ulysses' visit to the cave of the nymphs in Homer, *Odyss.* 13, 108–112. As a writer, Porphyry was distinguished by a clear insight into the meaning of the literary work he examined, and had an exceptionally lucid manner of stating that meaning. His *Isagoge* or introduction to the Categories of Aristotle was used for many centuries in east and west as the clearest and most practical manual of Aristotelian logic, indeed that logic was to a great extent popularized by the excellence of its presentation in the *Isagoge*. His " Sententiae " represent his exposition of Plotinus, again lucidly expressed but much preoccupied with his ethical teaching. He wrote a history of philosophy, of which his extant *Life of Pythagoras* no doubt formed a part. Like many neo-Platonists he was a vegetarian and ascete, which accorded with the tradition inherited from Pythagoras, as appears in the life of Apollonius of Tyana, a religious and moral reformer of the first century. One of his treatises, *De abstinentia*, deals with this ascetic ideal. He does not recommend abstinence from flesh for all, admitting that it is unsuitable for soldiers and athletes, but commends it to those who are occupied with philosophy : he disapproves the offering of animal victims in sacrifice, which he regards as a barbarous survival of the days when men had false ideas about the gods and as akin to human sacrifices which were obsolete since the days of Hadrian, animal sacrifices being in many cases a commutation of older human sacrifices. Animals have some measure of reason, and so have certain rights, they do not exist solely for the service of men. Abstinence from flesh food was practised by the Essenes, by the Egyptian priests, and by the Indian Sarmanoi, by which he denotes Buddhist priests about whom he obtained information from the Syrian Bar Daisan who had contact with an Indian embassy proceeding to Rome (Porphyry, *De abstinentia*, 4, 18). He repudiates the doctrine of transmigration of souls which to many people had made Pythagoreanism ridiculous. He was the author also of several works on psychology and mathematics.

Iamblichus (d. circ. 320), a native of Coele-Syria, was Porphyry's pupil in Rome and succeeded him as leader of the neo-Platonists. He was credited with supernatural powers, and it was said that at his devotions he was raised in the air and transfigured. His pupils asked him if this were true, and he laughed, and said that there was no truth in it whatever. As a writer he was inferior to Porphyry, with defects in style and often obscure, but the Emperor Julian considered him the equal of Plato, " a thinker who is inferior to him in time, but not in genius, I refer to Iamblichus of Chalcis " (Julian, *Orat.* 4, " On the Sun King," 146 A), and for some time, it appears, he had a great vogue. He wrote a treatise tracing philosophy back to Pythagoras, and of this some portions survive, including a life of Pythagoras. His *Logos Protreptikos* is an exhortation to philosophy which consists largely of extracts from Plato, Aristotle, and neo-Platonic writers. Besides these works he composed three mathematical treatises.

At the death of Iamblichus in 330, his school dispersed, but he had a successor in *Aedisius* at Pergamum in Mysia, who educated the sons of Eustathius, a noble Roman who was sent on an embassy to the Persian court. By that time the Roman Empire was professedly Christian, and the philosophers who adhered to paganism had to keep their religious sympathies secret. Amongst Aedisius' pupils was the Emperor Julian, who made an attempt to revive decaying paganism, but without permanent result. The great hope of the pagan party lay in the neo-Platonists. At the beginning of the fifth century Hypathia (d. 415) expounded neo-Platonic doctrines at Alexandria, but for the most part Alexandrian thought was not much attached to neo-Platonism. The same teaching was continued after her by Hierocles (circ. 415–450), a pupil of Plutarch of Athens (d. 481), who seems to have been responsible for introducing neo-Platonism into Athens which from his time forward became its home. Plutarch was succeeded at Athens by Syrianus of Alexandria. After him came Proclus (410–485) a native of Constantinople who received his education at Alexandria, then continued at Athens under Plutarch and Syrianus. He was the author of a treatise on " Platonic Theology " and of one called " Theological Elements ", which contains a statement of the doctrine of Plotinus modified in a form which supplied the philosophical ideas of the later neo-Platonists, so that he ranks next after Plotinus as an authority of their system. At that time the school of Athens, the home of neo-Platonism, was secretly pagan and conscious of the precarious character of the tolerance which it enjoyed. One of his pupils was Marinus, who wrote his biography.

The last head of the academy of Athens was *Damascius* a native of Damascus as his name denotes, but educated at Alexandria, then at Athens. He professed to accept the Aristotelian doctrine of the eternity of matter, in contradiction to the accepted Christian tenet of creation, and for this was viewed disapprovingly by the Emperor Justinian. But this was merely the climax of a growing antagonism of the imperial authorities for what was generally felt to be a nursery of paganism. Justinian's ideal was a centralized and united empire, in complete conformity with the ruling prince in religion and in everything else. Official disapproval led to a species of persecution of all philosophers in 528, and in the following year the school of Athens was closed and its endowments confiscated. Of the deprived professors seven, including Damascius, migrated to Persia and were welcomed by Khusraw, who was an ardent admirer of Greek philosophy and science. This migration seems to have taken place in 532. The seven philosophers expected to find an ideal state under the rule of a philosopher king, but were quickly disillusioned and discovered that an oriental tyranny could be worse than the severity of Justinian, and begged to be allowed to go back. Khusraw tried to induce them to remain, but used no compulsion, and when they did return took care to insert a clause in the treaty made with Justinian securing them complete liberty of conscience and freedom from molestation when under Roman rule. This return took place in 533.

Although the school of Athens was closed, the philosophers who had been trained there continued to teach and both they and their pupils produced written works. Chief amongst these late neo-Platonists were Ammonius and John Philoponus. *Ammonius* was a pupil of Proclus and compiled a commentary on the *Isagoge* of Porphyry which became the standard Greek authority and was afterwards adopted by the Nestorians. John Philoponus (*circ.* 530), a pupil of Ammonius, was a later commentator on the *Isagoge* and his exposition was preferred by the Monophysites.

(3) GREEK MATHEMATICIANS

The fame of *Euclid* (before 300 B.C.), one of the earliest scholars of Alexandria, did much to make the Museum a home of mathematical studies. His leading work, the *Elements*, probably contains a good deal which is not original, but is of great value as a summary of the knowledge of geometry acquired by the Greeks from the time of Pythagoras to his own days, arranged systematically and in logical sequence, a model method of statement, though more rigorous than is usual with modern mathematicians. Other works are attributed to him, some doubtful. Amongst them was a treatise on optics, probably apocryphal, which was used by the Arabs.

Aristarchus (d. *circ.* 230 B.C.), of Samos, the astronomer, was a teacher at Alexandria. He was the first to show how to find by means of the Pythagorean triangle the relative distances of sun and moon from the earth, though his result is not even approximately correct owing to the defective character of the instruments used. He also made the conjecture that the sun, not the earth, is the centre of the universe, a theory confirmed by Copernicus in the sixteenth century A.D. In this he does not seem to have had many followers, but his suggestion was not altogether forgotten and is mentioned by al-Biruni (*c.* A.D. 1000), who, however, did not adopt it.

Eratosthenes (d. *circ.* 194 B.C.) was a distinguished scholar of Alexandria and the leading geographer of antiquity. He devised a method of measuring the circumference and diameter of the earth, which was afterwards put into practice by the khalif al-Ma'mun in 829 and repeated a few years later. To do this he noted that at noon at Syene (Assouan) the sun was directly in the zenith, but at the same time in Alexandria it was 7° 12′ south of the zenith, and from this concluded that Alexandria was 7° 12′ north of Syene on the earth's surface. Knowing that the distance between the two places was 5,000 stadia, and as 7° 12′ is one-fiftieth of the full circle of 360° he calculated that the earth's circumference must be 50 by 5,000 stadia, i.e. 250,000 stadia, but altered that to 252,000 stadia so as to have 700 stadia exactly to a degree, thence computing its diameter to be equivalent to 7,850 miles of our measurement, and this is correct within fifty miles. He further stated that the distance between the tropics is eleven eighty-thirds of the circumference, making the obliquity of the ecliptic 23° 51′ 20″.

Archimedes (d. 212 B.C.), the friend of Eratosthenes, was not directly connected with Alexandria but his work, especially in mechanics, was known to and used by the Arabs.

Apollonius (*circ.* 225 B.C.), of Perga, was educated at Alexandria and applied himself to conic sections in which he used the names ellipse, parabola, and hyperbola. The work in which he dealt with this was in eight books, the first four of which are extant in Greek, the next three in an Arabic translation, and the last book is lost. The first four books, like Euclid's *Elements*, are a digest of material already known arranged in systematic order, books V to VII contain a good deal of new material due to his own research. He also composed other works on geometry.

Nicomedes (*circ.* 180 B.C.) was a writer of minor importance who is best known as the inventor of the conchoid curve by means of which an angle can be trisected.

Diocles (*circ.* 180 B.C.) invented the cissoid or " ivy shaped " curve which enables a cube to be duplicated, and studied the problem proposed by Archimedes of bisecting a sphere by a plane so that the volumes of the segments may be in a given ratio.

Hypsicles (*circ.* 180 B.C.), of Alexandria, may have been the author of what is known as the fourteenth book of Euclid, containing seven propositions on regular polyhedra. He also investigated polygonal numbers and certain indeterminate equations. In astronomy he introduced the division of the circle into 360 degrees and their subsequent sexagesimal divisions, though this he adopted from work already done by the Babylonian astronomers. The work of Hypsicles was translated into Arabic by Qusta b. Luqa, and afterwards revised by al-Kindi.

Hipparchus (d. *circ.* 125 B.C.) was not directly connected with Alexandria, but worked chiefly at Rhodes. He was the founder of scientific astronomy, which necessitated the measurement of angles and distances on a sphere, and in doing this he laid the foundations of spherical trigonometry. He worked out a table of chords, double sines of half the angle which was in use until the Indian system of calculating by sines was introduced by the Arabs. Plane trigonometry did not appear until later. He also made a catalogue of 850 fixed stars which marks the beginning of astronomy proper.

Heron (*circ.* A.D. 50), of Alexandria, was the inventor of several machines and wrote on dioptrics, mechanics, and pneumatics. Much of his mathematical work was concerned with the mensuration of land. He gives a formula for the sides of a triangle which may be represented as :—

$$\triangle = \sqrt{\{s(s - a)(s - b)(s - c)\}}$$

where s = a + b + c.

In his geometry appears the rule which we express as—

$$c = \frac{n}{4} \times \cot. \frac{180°}{11}$$

where n = number of sides of a polygon of area A and side s,

and where $c = \frac{A}{S^2}$.

He was able to solve the equations which we represent as

$$ax^2 + bx = c.$$

Heron was translated into Arabic by Qusta b. Luqa (mechanics).

Menelaus (*circ.* A.D. 100) wrote on the sphere and spherical triangles, also six books on calculating chords. He states the theorem that if the three sides of a triangle are cut by a transversal, the product of the lengths of three segments which have no common extremity is equal to the products of the other three. Menelaus was not directly connected with Alexandria, but is known to have taken astronomical observations in Rome.

Nicomachus (*circ.* A.D. 100) also had no direct connection with Alexandria. He wrote a treatise on music and two books on arithmetic, possibly a compendium of a larger work now lost.

Marinus (*circ.* 100 A.S.), of Tyre, was a geographer who improved on the methods of Hipparchus. He located places by the use of two co-ordinates, latitude and longitude, but his work has not come down to us, most of it no doubt incorporated in that of Ptolemy.

Claudius Ptolemy (*circ.* A.D. 140–160) taught both in Athens and Alexandria. His chief work was known as the Μαθηματικῆς συντάξεως βιβλίον πρῶτον. He wrote another σύνταξις and therefore the Arabs called the principal treatise ἡ μεγίστη and placing the Arabic article before the name made it *almajest*. He gives a summary of all earlier work on the size of the earth and the exact position of certain places. He further developed Hipparchus' table of chords and extended the use of sexagesimal fractions. His work in astronomy has been justly compared with that of Euclid in geometry, it gave an ordered and logical summary of all that had been done so far. He increased Hipparchus' catalogue of 850 fixed stars to 1,022. In astronomy he took the earth as the centre of the universe and planned a complicated system of cycles, eccentrics, and epicycles to account for the movements of the heavenly bodies. This system apparently held good to a certain point, then it was detected to be unsatisfactory by Arab astronomers and efforts were made to correct it, the best known being that of the " new astronomy " which arose in Andalus (Arab Spain) in the eleventh century, but no correction produced a completely satisfactory result until the whole was completely replanned after Copernicus proved that the sun is the centre of our system and that the earth and other planets revolve around it. He was also the author of a work on astrology, the *Tetrabiblos*, which had a good deal of influence over Arab thought. A good deal of his work was translated into Arabic by Yusuf al-Hajjaj, the *Tetrabiblos* by Abu Yahya al-Batriq, whilst his geography formed the basis of al-Khwarizmi's *Book of the Image of the Earth* which reproduced his maps in a modified form.

Diophantus (*circ.* A.D. 250), of Alexandria, was the author of an arithmetic in thirteen books of which six survive, a treatise on polygon numbers of which part is extant, and a collection of propositions which he called porisms. The first of these deals with the theory of numbers and includes an algebraical treatment of arithmetical problems. In solving determinate equations he recognized only one root, even when both roots are positive. He treats also some indeterminate equations and certain cases of simultaneous equations. He did not exactly invent algebra, but prepared the way for it by a treatment of arithmetic which anticipated algebra. His work influenced both Indian and Arab mathematicians, but neither followed him with sufficient confidence to make full use of the path he opened. It was not until the rediscovery of his work in sixteenth century Europe that full advantage was taken of his methods and so a foundation was laid of modern algebra.

Pappus (*circ.* 300), of Alexandria, wrote eight " books of mathematical collections ", of which the first two are lost, but the remaining six are extant. Of these six, Book III deals with proportion, inscribed solids, and duplication of the cube ; Book IV, spirals and other plane curves ; Book V, maximum and isoperimetric figures ; Book VI, the sphere ; Book VII, analysis ; and Book VIII, mechanics.

Hypatia (d. 415), of Alexandria, daughter of the mathematician Theon, is said to have written a commentary on an astronomical table of Diophantus, possibly not the distinguished mathematician already mentioned, and on the conics of Apollonius, but neither of these survive.

Proclus (d. 485) studied at Alexandria and taught at Athens. He wrote many books, including a paraphrase of portions of Ptolemy, a work on astrology, another on astronomy, and a commentary on the first book of Euclid's *Elements*.

(4) GREEK MEDICINE

The history of Greek medicine proper begins with Hippocrates, of Cos, who died in 257 B.C., and his "Aphorisms" always remained a leading text-book for practitioners. This collection of aphorisma was amongst the early medical works translated into Arabic by Hunayn ibn Ishaq, who was able to use the Greek text. There is an anonymous Syriac translation which has been published by Pognon (Leipzig, 1903), but its date does not appear.

In the later period of the school of Alexandria the medical works of Galen (d. A.D. 200) were established as the recognized authority, and a selection of his treatises formed the official curriculum for medical study. This curriculum was reproduced at Emesa and Jundi-Shapur and Syriac versions were prepared for the use of Syriac-speaking students. Many of those Syriac translations were made by Sergius of Rashayn, but were afterwards revised by Hunayn ibn Ishaq and his companions in the Dar al-Hikhma at Baghdad, or were supplanted by new versions prepared at that academy. This translation into Syriac preceded the preparation of Arabic versions, but went on for some time side by side with translation into Arabic. Galen himself had practised at Rome, but his studies were made at Smyrna, Corinth, and Alexandria.

The chief Greek medical writers after Galen were :—

Oribasius (born *circ.* 325) was a friend of the Emperor Julian and the person whom Julian selected to be the confidant of his dissatisfaction with Christianity and determination to revert to paganism. This letter (Julian, *Epist.*, xvii) was probably written in 358. He was with Julian in Gaul and accompanied that prince's unfortunate expedition into Persia where he was present at his death in 363. After his return from Persia his property was confiscated by Valentinian and Valens, though the reason for this is not clear. He was then banished to a " land of barbarians ", but this could not have been for long as he returned in 369. Three of his medical works are extant, one of these was a *Synopsis* dedicated to his son Eustathius in nine books, and this was translated into Arabic by Hunayn ibn Ishaq and was known to 'Ali 'Abbas. It is quoted by Paul of Aegina.

Aetius (end of the fifth century) was a physician who practised at Constantinople. Nothing is known of his life, even the date of his activity is unknown, but he is supposed to have lived in the later fifth century as he refers to Cyril of Alexandria, who died in A.D. 444 and to Petrus Archiater who was physician to Theodoric, King of the East Goths. He was a Syrian of Amida. He was the author of a medical compendium in sixteen books, now divided into four groups. His work does not contain much original matter, but its contents are well chosen. He was the first Greek physician to give serious attention to spells and incantations.

Paul of Aegina, probably of the late seventh century. Nothing is known of his life. Suidas says that he was the author of several medical works. Of such works one only is extant and is known as *The Seven Books on Medicine*. This was translated by Hunayn ibn Ishaq and was in great repute amongst the Arabs, especially as an authority on obstetrics, for which reason he was surnamed *al-qawabil* " the accoucheur " by them.

Aaron, priest and physician, of Alexandria, is another about whose life no information is available. He was the author of a *Pandects* or *Syntagma*, which is said to have been translated into Syriac by a certain Gosius. This Gosius has been identified with Gesius Petaeus who lived in the days of the Emperor Zeno (474–491). The late Syriac writer Bar Hebraeus states that Aaron composed thirty books which were translated by Sergius, of Rashayn, who added another two books, but Steinschneider holds that these additional books were the work of the translator who made the Arabic version, a Persian Jew named Mesirgoyah. Aaron's works circulated amongst the Arabs and had a considerable influence on Arab medicine.

CHAPTER IV

CHRISTIANITY AS A HELLENIZING FORCE

(1) HELLENISTIC ATMOSPHERE OF CHRISTIANITY

THE Christian Church in its earlier period was essentially a Hellenizing force. Its language was Greek and its first outspread was amongst those who were Greek in speech and culture, if not in race. Even in Rome itself it used Greek, as appears from the fact that the early Christian Roman writers, Clement, Hermas, Hippolytus, and others wrote in Greek. Greek is the language generally used in the earlier catacomb inscriptions, and seems to have been that employed in the primitive Roman liturgy, though the Greek phrases now surviving in that liturgy were added later, probably in the fifth century, the *Kyrie eleison* introduced by St. Gregory at a still later date (John the Deacon, *Vita S. Gregorii*, ii. 20, *P.L.* lxxv, 94). This prevailed until well into the fourth century when Constantine removed the imperial government to New Rome (Constantinople). The churches of Gaul also were Greek-speaking, though not to so late a period, and the province of Africa, afterwards the home of Latin Christianity, seems to have had a primitive Greek phase, if Aubé is right in regarding the Greek text of the Acts of the Martyrs of Scillite discovered by Uesener in 1881 as the original (Aubé, *Etude sur un nouveau texte des actes des Martyrs Scillitains*, Paris, 1881) : Greek seems to have been largely used in second century Carthage. All this shows that Christianity spread first through the urban commercial population round the Mediterranean whose lingua franca was Greek. It was only later that it penetrated into the hinterland and reached the vernacular-speaking populations of Egypt, Syria, Italy, Gaul, and Africa. Greek was an international language and Christianity appeared as an international religion.

It is of course true that Christianity claimed a Jewish origin, for " salvation is of the Jews " (St. John iv, 22), but it developed in an atmosphere of Hellenistic Judaism, such as produced Philo of Alexandria, who used his Old Testament in Greek, not in Hebrew.

The Diaspora or Dispersion of the Jews began after the destruction of Jerusalem by the Babylonians in 588 B.C., when many of them found a refuge in Egypt. The Babylonians were conquered by the Persians under Cyrus in 538, and Cyrus permitted the rebuilding of Jerusalem and the restoration of its temple. But many of the Jews who had migrated to other lands did not want to go back to Palestine, finding much better openings elsewhere, and this was especially the case with those who had gone to Egypt, where they had formed several populous and flourishing colonies. When Alexander founded Alexandria in 332 he invited Jews to his new city and assigned them one out of the three regions into which it divided (Josephus, c. Apionem, 2.4 ; Bell. Jud. 2.18.7). These Egyptian Jews, however, formed an integral part of the Jewish community, recognized the jurisdiction of the High Priests, and paid regular tribute to the temple at Jerusalem. Although under the rule of the Seleucid monarchs of Syria, they retained their own laws and religion without interference to the reign of Antiochus Epiphanes (175–164 B.C.), who began trying to Hellenize them and to introduce the worship of Greek deities in Jerusalem. This resulted in a revolt led by the Maccabees which Antiochus was unable to put down. At the beginning of his reign Antiochus deposed the High Priest Onias III and put his brother Jason in his place, then substituted a younger brother Menelaus or Onias IV, who procured the murder of Onias III. Onias V, the son of the murdered ex-High Priest, fled to Egypt to escape the sacrilege and disorder produced by Antiochus' policy and with him went some adherents who esteemed him to be the legitimate High Priest. They were well received by Ptolemy Philometor (181–146 B.C.), who gave them a disused Egyptian temple at Leontopolis, and there they constructed a replica of the temple at Jerusalem and duly observed the daily sacrifices and other rites. This temple at Leontopolis remained in use until the temple at Jerusalem was destroyed in A.D. 70, and then it was closed. Although a sanctuary of the Egyptian Jews this local temple never attained the prestige of the temple at Jerusalem, to which tribute was sent from Egypt as from other countries of the dispersion. Probably it was in connection with this temple that a Greek translation of the Old Testament, known as the Septuagint, was made, apparently by gradual stages, the translation of the five books of Moses in a rather crude vernacular such as was used in Egypt and which has its parallel in many of the Egyptian papyri, and this translation was made early enough to be used by Demetrius (as cited in Clemens Alex., *Stom.*, i, 21, and Eusebius, *Praep. Evang.*, ix, 21, 29), who probably lived under Ptolemy Philopator (222–205), whilst the historical and prophetical books were translated later in more literary form, and the latest books, Ecclesiastes and Song, in an improved and more literal style. The legend of Seventy Elders who made the translation under Ptolemy Philadelphus (285–247 B.C.), based on the spurious letter of Aristeas to his brother Philocrates, is unhistorical. Probably the whole translation was not completed before the early years of the Christian era. Philo of Alexandria does not quote from Ruth, Ecclesiastes, Song, Esther, Lamentations, Ezekiel, or Daniel, nor does the New Testament quote from Ezra, Nehemiah, Esther, Ecclesiastes, Song, or certain of the Minor Prophets.

Beginning with the revolt of the Maccabees there was a strong anti-Hellenist reaction in Palestine which seems to have spread abroad amongst the Jews of the Dispersion in the early years of the Christian era. It was part of the nationalist movement which inspired the Jewish revolt that culminated in the destruction of Jerusalem. This reaction returned to stricter observance of Hebrew tradition, to the use of the Hebrew language, and to the older idea of complete separation from the " gentiles ". This reaction was the parent of Rabbinical Judaism. In this stricter Judaism it was no longer tolerated to read the scriptures publicly in the synagogue in the Greek language, the observance of the rite of circumcision and all other legal ordinances was punctiliously enforced and any familiar intercourse with pagans or the " uncircumcised " was absolutely forbidden. The Mosaic law was made stricter by rabbinical glosses.

The rivalry between this stricter traditional party and the laxer Hellenistic Jews of the Dispersion had its repercussion in the Christian community. There were at first two parties, Judaistic Christians who wanted all converts to be circumcised and subject to the whole Mosaic law, and Hellenistic converts who demanded no more than the acceptance of the Christian faith. The controversy between these two parties is recorded in the Acts of the Apostles. In the end the Judaistic party disappeared altogether, for the Judaistic Christians which appear later in the Antioch of St. John Chrysostom belonged to a heretical sect which deliberately tried to revive Jewish usages. Possibly it may be said that Christianity is the heir of Hellenistic Judaism, the inheritor of that monotheistic moral religion which so well suited the trend of Hellenistic thought.

The Christian Church received the Old Testament, but used it as subordinate to the New. The prophecies were treated as referring to Christ, its moral teaching as preparatory to a fuller revelation in the gospel. As the Greek converts greatly outnumbered the Jewish ones, it is not surprising that Greek education, which implied Greek philosophy, very soon began to permeate Christian teaching. Indeed it had already influenced Jewish thought as can be seen in several books of the apocrypha, such as Wisdom and Ecclesiasticus, which bear the impress of Stoic thought. In this, as in many other respects, Christianity only continued the logical evolution of Hellenistic Judaism. In this adaptation of Christianity to gentile thought the leader was St. Paul whose epistles had a great influence on the formation of Christian doctrine and its approximation to current Greek philosophy. Like the Hellenistic Jews the Christians used the Old Testament only in its Greek version, and the earlier formulation of its doctrine was expressed in terms borrowed from Greek philosophy. Thus from the beginning the Christian Church was shaped to be the teacher of Greek intellectual culture as well as of evangelical doctrine. Later, when controversies arose within the Church, these too were expressed in Greek philosophical terms and fought out according to philosophical principles.

Religion may be concerned only with ritual, which is the case with most primitive religions, concerned only with sacrifices and the due performance of sacred rites. A later stage is reached when religion becomes a moral agency, which begins perhaps with the observance of tabus. Last comes the development of speculative theology, itself a form of philosophy which seeks to explain why things are as they are and to account for man's place in the universe. The ancient Egyptian religion seems to have reached this final stage in its later days, but in Greek thought philosophy had superseded or absorbed religion, and it was in a society where philosophy had practically replaced religion that Christianity was evolved. The old Greek and Roman religions, purely ritual and very largely magic, had no living influence and held their ground only as traditional survivals to which people were attached by long association. Morality was absorbed in philosophy as well as speculation on man's place in the universe, indeed his duty was essentially involved in the reason for his existence. Thus Christianity was presented rather as a philosophy which set itself to unravel the problem of existence. Undoubtedly it borrowed a good deal from the mystery religions with which it had certain similarities, but the dominating influence in the evolution of Christianity was the current attitude of the Hellenistic world towards religion, which was a philosophical attitude. In fact philosophy had replaced religion in the older sense.

Although the Church inherited the Jewish scriptures and followed the synagogue precedent in its liturgy, it definitely broke with Judaism, and the break was clearly seen by the Jewish authorities. Judaism was reverting to the ritualism of the past and to national exclusiveness; Christianity advanced into a freer and more open atmosphere for which Alexander's Conquests had cleared the way. It was a centrifugal movement, Judaism going farther towards the right, Christianity towards the left. The Jews aimed at a reformation by complete reversion to the past, which always is the professed aim of religious reformation. They regarded the Christians with aversion as pressing on more recklessly on the path of laxity which they esteemed the cause of their own decadence. At a later period Jewish philosophers and scientists made a valuable contribution to intellectual culture, but that was in days when they were under Arab rule. No such tendency appears in the older Jewish academies of Sora and Pumbaditha where interest was concentrated in law and ritual observances.

(2) Expansion of Christianity

The early Church, as pictured in the Acts of the Apostles and the epistles of St. Paul, undoubtedly had a missionary spirit. But that missionary spirit first appears as resulting from persecution. It is related that the first " scattering " of

Christian teachers from Jerusalem took place when persecution followed the martyrdom of St. Stephen. Very often in after times a similar reason led to the preaching of Christianity in new districts. Probably the British Church owed its origin to refugees from the persecution which broke out in Lyons and Vienna. Persecution was not the only cause of the outspread of Christianity, but it was one cause, and perhaps a leading one.

Jewish opposition appears plainly in the narrative of the Acts, and Jewish antagonism seems to have been the principal cause of many, but not all, the earlier persecutions of the Church. The first actual persecution of Christians as a community took place in Rome under Nero, certainly instigated by Jews who were powerful at court. After this there were outbreaks of popular antagonism in many parts, especially in Asia Minor where Christians were numerous, and in some of these outbreaks Jewish influence seems to have been active. Under Trajan some attempt was made to regularize the policy to be followed in dealing with the Christians. When Pliny was governor of Bithynia he found many Christians there and a good many disturbances took place for which they were blamed. Pliny had had experience of legal administration in Rome, but apparently had had no contact with cases connected with Christians, as such cases came before the Praefectus Urbis or his deputy. He sought the Emperor's guidance, and Trajan replied in letters which gave a precedent for dealing with persons charged with practising this unauthorized religion. It was decided that Christianity was a crime deserving of death, but it was not permitted to make search for Christians and informers against them incurred penalties. At a later period Domitius Ultianus compiled a treatise, *De officio proconsulis*, of which the seventh book gave a summary of anti-Christian legislation. This work would have given us a complete view of the attitude of Roman law towards the Christians, but unfortunately only a few extracts survive, the most important is Lactantius' indignant criticism (Lactantius, *Instit.*, v, 11, 12). The subject remains obscure, which is to be regretted as undoubtedly persecution, or at least liability to persecution, was a strong motive causing Christians to go outside the Roman Empire, and so one of the chief causes of the spread of Christianity.

Some light is given by Hippolytus' account of Callistus,

a Christian slave who was entrusted by his master, also a Christian, with funds to open a bank, but went bankrupt. He tried to recover loans from debtors, amongst them some Jews, and was alleged to have disturbed a synagogue in his efforts to get hold of them, and for thus disturbing the worship of a legally authorized community was brought before a judge. Obviously the Jews worked hard to get him accused of Christianity by bringing this out incidentally in the evidence : they could not bring it as a direct charge for fear of incurring the penalties attached to laying information. Callistus was sentenced as a Christian and condemned to labour in the Sardinian mines, but after some time was included in a pardon obtained by Marcia, the concubine of the Emperor Commodus, who either was herself a Christian or very well disposed towards the Christians. (Whole incident in Von Döllinger, *Hippolytus und Kallistus*, ch. viii.) All through the third century Christian interest was strong at court (cf. Eusebius, *H.E.*, vi, 34 ; vii, 10). The effective cause of the violent but brief persecutions under Decius and Diocletian towards the end of that century was that the Christians had become too powerful, practising their religion too openly and building large churches. Before Decius they had been protected by Roman law in holding property and the subterranean cemeteries of Rome, covering considerable areas, were their acknowledged property from the time of Pope Zephyrinus (202–219) : it was an innovation when Decius tracked down Christians even in their cemeteries and seized their property. Persecution was occasional and spasmodic, usually provoked by non-religious motives, but there was a liability to persecution, and this undoubtedly led to some Christians going outside the Roman frontiers, or at least moving to a province where persecution was comparatively rare. The first beginnings of the British Church seem to have been due to fugitives from persecution in Gaul, and that church was by no means the only one which traced its origin to refugees.

The desire to be safe from the liability to persecution seems to have been responsible for the formation of a flourishing church in Mesopotamia outside the Roman Empire. This Mesopotamian Church, chiefly about Edessa, lived its own life in a comparatively free atmosphere, and developed its own style of church building and, apparently, its own system

of discipline.   Later, when the empire became Christian and
the Catholic Church was directed by Greek bishops, much
of this local Mesopotamian development was suppressed with
a high hand, but the fact remains that some of the earliest
extant evidence of church organization and building belongs
to the area just across the eastern frontier of the Roman
Empire.   This Mesopotamian area had experienced Greek
influence under the Seleucids.   Greek influence was brought
to bear by the Romans whose frontier towards Parthia swayed
back and forth from time to time and who always had political
interest in the border lands.   But it was the Church more than
anything else which brought about the Hellenization of that
area across the frontier.

As it grew in prosperity the Church produced literature.
In Alexandria, as might be expected, some of its earliest
writers appeared, Clement of Alexandria, Origen, and others,
and about A.D. 180 Hegesippus travelled about the
Mediterranean world investigating evidence for the apostolical
tradition of the Church's teaching and institutions.   Shortly
before his time Justin Martyr shows a Christian teacher trying
to combine current philosophy and Christian doctrine.   By
the end of the second century Christianity was not merely
strong in the number of its adherents but well reinforced by
its literary output and its co-operation with philosophy.
Christian literature was in Greek, the earliest vernacular
Christian literature which came after was produced in Syriac
and its classical standard was the dialect of Edessa, much
earlier than any Christian material in Latin.   Throughout
the Church generally the Old Testament was known only in
its Greek translation, as had been the case with the Egyptian
Jews in the days of Philo of Alexandria, and presumably
with the Hellenistic Jews generally.   Vernacular versions of
the Old Testament are mostly translated from the Greek
Septuagint, the older Syriac version alone shows an independent
source which is closer to the Hebrew original.   It may well be,
however, that the Masoretic text which became the authorized
version of the Old Testament represents a text selected from
earlier divergent and varied texts, so that the Septuagint
and its versions sometimes at least go back to an older form
which has been rendered obsolete in Hebrew by the acceptance
of a standardized text.

(3) ECCLESIASTICAL ORGANIZATION

Although the Christian Church traced its origin from the Jewish synagogue, it appears in history in a structure organized, not on Jewish lines, but on lines following the structure of the Roman Empire. This began before the Church had received formal toleration, but became more pronounced after toleration had brought the Church into closer relations with the secular State. It was in 313 that the Emperor Constantine granted formal toleration to the Christian religion and in 325 summoned the first general council at Nicaea to define disputed points in Christian doctrine and regulate discipline. From that time forward the Church was protected and to some extent controlled by the State, though it was not until the days of Gratian (368) that it was recognized as the established religion.

In its earlier days the Church consisted mainly of urban congregations, over each a bishop with supporting group of presbyters. But gradually it spread out into the rural areas and congregations were added in outlying parts with presbyters only, each attached in discipline to a neighbouring bishop. Thus territorial dioceses were formed as the Church expanded from the cities which had been its earlier home. Already in Nicene times these territorial units were gathered together into confederations, like civil provinces, each known as a diocese, the name having a much wider scope than it now possesses. In the Eastern Church there were four such dioceses, the Orient, Pontus, Asia, and Thrace. These were divided into eparchies, each with one or two metropolitans. Thus Asia comprised the eparchies of Ephesus, Sardis, Smyrna, and Pergamum. The chief bishop or metropolitan of each eparchy came to be known as an archbishop. In the end there was a general recognition of the primacy of the great churches of Rome, Antioch, and after some hesitation Alexandria. Afterwards for sentimental reasons Jerusalem was conceded similar rank, though in fact subordinate to Antioch. The council of Chalcedon (canon 28) terminated the independence of Pontus, Asia, and Thrace and put them under the bishop of Constantinople which was thus raised, in spite of protests, to equality with Antioch and Alexandria. The bishop of these greater groups of churches was called patriarch, a name in frequent use in the post-Nicene age, but not formally recognized by any conciliar decree until the ninth century.

The Mesopotamian Church across the frontier was regarded as within the diocese of Antioch, but at an early date its chief bishop received the title of Catholicus, a title already employed by Constantine in writing to the Bishop of Carthage, and used in the civil administration for a procurator or deputy of a provincial governor. This title is used by Procopius (ii, 25) for the head of the Persian Church and ultimately became the perquisite of the Bishop of Seleucia. After the Nestorian schism the bishops of Seleucia appropriated it as the distinctive title of the head of the Nestorian community.

From the Nicene age onwards the Church was steadily organizing itself on lines similar to those already employed in the civil administration of the empire, though the areas of provinces, dioceses, and eparchies was not in all cases identical with those of the civil structure. Thus organized as a kind of replica of the Roman Empire it very efficiently and thoroughly assimilated the Christian communities, not only of Mesopotamia but also of Persia, to Hellenistic standards. Such standards applied to social organization prepared the way for Greek culture. The Christian religion, unlike some of the older religions, was not based on ritual observances alone, nor entirely on rules of moral conduct. The Greek influence it inherited came from that later Greek thought in which religion was absorbed in philosophy. Christianity set a body of theological doctrine in the forefront : ritual observances were designed as expressions of that body of doctrine, and morality also was built up on a basis of doctrinal teaching. All this doctrine was strongly coloured by philosophy, much of it was simply philosophy expressed in theological terms. The philosophy thus adopted and utilized by the Christian Church was that philosophical teaching current in the Greek world during the earlier centuries of the Christian era, the eclectic philosophy which professed to be derived from Plato and Aristotle. Such philosophy guided and directed the controversies raised in the Church by Arius, Nestorius, Eutyches, and others. The problems debated were suggested by philosophy, the conclusions reached were the results of philosophical treatment. Perhaps the most salient point is the complete adoption of the Aristotelian logic as the means of investigation and argument. However much Christian sects differed in their tenets, all alike accepted the Aristotelian logic as the method to be employed in investigation and solution.

Thus the Christian Church remodelled the communities of its converts in conformity with the social structure of the Roman Empire, grouping Persians, Arabs, and other orientals according to a system of dioceses and provinces which was copied from the imperial administration, and promulgated amongst them educational standards which reproduced those established in Alexandria. The chief source of scientific and philosophical material received by the Arabs came through Christian influence.

It has been disputed whether Muhammad owed most to Jewish or Christian predecessors, apparently he owed a great deal to both. But when we come to the 'Abbasid period when Greek literature and science began to tell upon Arabic thought, there can be no further question. The heritage of Greece was passed on by the Christian Church.

CHAPTER V

THE NESTORIANS

(1) THE FIRST SCHOOL OF NISIBIS

NISIBIS lay within the territory ceded to Rome in 298. As it then became a frontier town commanding the main route between Upper Mesopotamia and Damascus, the Romans fortified it very strongly. Probably there already were Christians there, as in so many parts of Mesopotamia, and some few years later, in 300 or 301, it was recognized as an episcopal see, its first bishop Babu, who was succeeded by Jacob. The town had a great many Jewish inhabitants also and possessed a Jewish academy founded by R. Judah ben Bathyra, an eminent *tanna* seventeen of whose *halakoth* are quoted in the Mishna. Probably there were three persons of this name, father, son, and grandson : the first living whilst the Temple was still standing in Jerusalem, the last contemporary with R. Akiba, with whom he is said to have had controversies. Apparently the Jews suffered severely when the Romans took the town, and it is probable that this involved the end of their academy, at any rate it is not mentioned afterwards.

Bishop Jacob attended the Council of Nicaea in 325 and subscribed its decrees. Not long after that council Eustathius, Bishop of Antioch, founded a school at Antioch in imitation of the great school of Alexandria, and his example was followed by Bishop Jacob who founded a similar school at Nisibis, with the special purpose of spreading Greek theology amongst Syriac-speaking Christians, whose theology and the arrangement of whose churches, as Strzygowski points out, did not conform to the accepted standards of the Catholic Church. He placed a presbyter named Ephraem in charge of this academy. Ephraem became a celebrated teacher and raised the school of Nisibis to great fame. Not only so, but he was also distinguished by his literary work. He was not the first to write in Syriac, but in later ages he was always regarded as the standard authority for classical Syriac. Whilst he presided over the school at Nisibis he composed poems which became the models of Syriac verse. He is said to have presided over the school for a period not far short of sixty years, presumably he was quite a young man when he was appointed, and the end of the school was by no means the end of his career. The chronology, however, is not altogether clear.

The school at Antioch had a chequered history. Comparatively early in its career, in 331, Eustathius himself was sent into exile and left the school in the hands of Flavian, who took as his associate Diodorus, an ascete who had long been his intimate friend. All these three, Bishop Eustasius, Flavian, and Diodorus were prominent in controversy with the Arians, a prominence responsible for many of the troubles which came upon the school of Antioch, for at the time the Arians had much political power, and that became more so after the death of Constantine in 337. The school, however, continued until 379 when Diodorus became Bishop of Tarsus : in 381 he was one of the bishops who consecrated Flavian to the see of Antioch. When Diodorus was raised to the episcopate the school dispersed, but one of its teachers, named Theodore, continued teaching a few members who adhered to him until 392, when he was made Bishop of Mopseustia. Diodorus of Tarsus and Theodore of Mopseustia came to be regarded as the leading theologians of the Syrian Church, the Greek-speaking church dependent on Antioch, and their writings which, of course, were in Greek, were taken as the bulwarks of the faith in Syria. Greatly revered as teachers of orthodoxy their teaching differed in method from that in vogue in the school of Alexandria, and it would seem that such difference in scholastic method was accentuated by a racial jealousy between the Syrians and Egyptians, for certainly there was a rivalry, not altogether friendly, between Antioch and Alexandria. No one could have suggested any doubt as to the orthodoxy of these two distinguished theologians, but later ages suspected them of having sown unintentionally the seeds of Nestorianism, and some incautious expressions used by Theodore were seized upon as suspect of implied Nestorianism and so were formally condemned at the fifth General Council held at Constantinople in 553.

Meanwhile Nisibis also had its troubles. Bishop Jacob died probably soon after 341 when he was visited by Milles, Bishop of Susa in Persia. Not long after this came Julian's unfortunate expedition against Persia, and after its disastrous end in 363 the five provinces acquired by the Romans in 298 had to be handed back to Persia. In the war which ended thus calamitously Ephraem, the head of the school of Nisibis, had taken a leading part in defending the city against the Persians and, as the city now passed into Persian occupation, he felt it impossible to remain there and fled to Edessa.

No doubt there were many other refugees and Ephraem, as an unknown fugitive, had to undertake menial labour to earn his daily bread. For some time at least he found employment as an attendant in the public baths. But friends discovered him and encouraged him to resume teaching, and thus a Christian school was established at Edessa. The school of Nisibis had not migrated to Edessa, it had simply scattered when Nisibis fell into the hands of the Persians, but as its head resumed his work in Edessa there was a continuity between the two schools and that of Edessa may be considered as a revival of the school of Nisibis. Ephraem lived twelve years after the fall of Nisibis and died in 375. Not all that period was spent in teaching; besides his literary work he seems to have travelled and to have spent some time as a hermit. After his death the school had a prosperous career. Its teaching was carried on in Syriac, the Syriac of Edessa being reckoned as the literary dialect of Syrian Christians.

In 412 Rabbula was appointed Bishop of Edessa. He was the son of a converted pagan priest of Kennesrin (Chalcis) and a man of considerable energy. The school was under a teacher named Ihihba or Hibha, whose name is rendered in Greek as Ibas. Some while before this there had been a revival of learning which seems to have commenced in Asia Minor, probably in Cappadocia, and reached the Syriac-speaking community in the course of the fifth century. It seems to have been connected with an ecclesiastical development which centred at Caesarea in Cappadocia. From St. Gregory Thaumaturgus and onwards the church there attained a great reputation as a model in matters liturgical (cf. Brightman, *Eastern Liturgies*, Appendix N, pp. 521–8), which culminated in a revised liturgy produced by St. Basil (d. 379), which became the established rite of Constantinople and still remains the principal liturgy of the Orthodox Greek Church. The second Greek liturgy, in more general use, bearing the name of St. John Chrysostom (d. 407), is simply an abridged form of the liturgy of St. Basil, whilst there is a third form, wrongly ascribed to St. Gregory (d. 604), which also is based on St. Basil. Of these the full liturgy of St. Basil is now used only on the Sundays of Lent (except Palm Sunday), Maunday Thursday, the eves of Christmas, Epiphany, and Easter, and on St. Basil's day (1st January) : that of St. Gregory is used on weekdays in Lent. This liturgical development was a by-product of an extensive and influential wave of cultural influence which spread out from Cappadocia to Byzantium, and then passed onwards through the Oriental churches into Asia. Edessa, as the metropolis of the Syriac-speaking Church and the focus of the Syriac phase of Hellenistic intellectual life, became the distributing centre of the Cappadocian renaissance.

(2) SCHOOL OF EDESSA

Nisibis was taken by the Persians in 363, and Ephraem, who had been its head, fled to Edessa. As a refugee he had to earn his livelihood in a humble way and entered the service of a bath-keeper, but devoted his spare time to teaching and reasoning with those who cared for his company. One day when he was thus occupied he·was overheard by an aged hermit who had come down from his hermitage to visit the city, and who rebuked him for being still interested in earthly knowledge. This caused Ephraem to retire to the mountains and spend some time in a hermitage meditating, reading, and literary composition, which bore fruit in some of his hymns and poems. At that time a revival of learning which greatly influenced the Church was in progress in Cappadocia, especially associated with Basil of Caesarea, and this induced Ephraem to travel to Cappadocia and visit Basil, perhaps going to Egypt, the " holy land " of monasticism, on the way. Before long, however, the news that Edessa was disturbed by the teaching of various heresies arising out of the teaching of Bar Daisan who had lived in that city in the second century, caused him to return and resume his teaching. Later he again retired to the hermit life, but this time was recalled by the news that Edessa was suffering from a severe famine, and by his presence and exhortations he succeeded in inducing the wealthier citizens to give generously to the relief of their more indigent neighbours. His death took place not long afterwards in 373. Considering these interruptions of the ten years of his sojourn in Edessa we can hardly regard him as organizing and directing a school there, but it appears that his influence gave impetus and direction to the group of disciples who gathered round him, and after his visit to Cappadocia this meant that they were brought into touch with the Cappadocian renaissance.

Ephraem's most prominent pupil was Zenobius Gaziraeus, a deacon of Edessa, who wrote against the Marcionites and was the teacher of Isaac of Antioch. At first the school of Edessa seems to have been an informal group, so that it is hardly possible to describe Ephraem as its first head and Zenobius as his successor. But out of this group gradually developed what became a well-known academy, though it had no official and formal foundation like the schools of Nisibis and Antioch. It might, of course, be reckoned as a continuation of the school of Nisibis closed in 363, as it was commenced and guided by one who had been the official head of the Nisibis school, but there was no migration of teachers and students which could justify its being regarded as a colony of Nisibis.

There is plain evidence of work done at Edessa in the later fourth century in translation from Greek into Syriac. The manuscript, Brit. Mus. Add. 12150 of date 411, contains Syriac translations of the *Theophania* and *Martyrs of Palestine* of Eusebius, and of Titus of Bostra's discourses against the Manichaeans, whilst a St. Petersburg manuscript of 462 contains a Syriac version of the *Ecclesiastical History* of Eusebius. (The Syriac version of the *Theophania*, edited by S. Lee, London, 1842, trans. Camb., 1843; of the *Martyrs of Pal.*, ed. trs. W. Cureton, London, 1861; of the *Eccles. Hist.*, by W. Wright and N. McLean, Camb., 1898; of *Titus of Bostra*, P. de Lagarde, Berlin, 1859.) Internal evidence shows that these texts have passed through the hands of a succession of scribes, so must have been made some time before 411 and 462 respectively. Eusebius died in 340, Titus of Bostra in 371, so the translations into Syriac may have been made during the authors' lifetime, or very shortly afterwards, as was the case with the letter of Cyril, of Alexandria, " On the true faith in our Lord Jesus Christ to the Emperor Theodosius," which Rabbula, the Bishop of Edessa, translated into Syriac as soon as he received a copy from its author.

The school was well established and of good repute amongst the Syriac-speaking community of Mesopotamia and Persia and most of Persian bishops were its alumni when in 411–12 Rabbula was appointed Bishop of Edessa, and about the same time or soon afterwards *Hibha* (Ibas) was made head of the school. The works of Theodore of Mopseustia, and Diodorus of Tarsus, were then the standard theological authorities of the Syrian Church, and Hibha made a Syriac version of Theodore's work for use at Edessa and then, as the terminology and logic of that work offered difficulties to oriental students, he also made a Syriac translation of the *Isagoge* of Porphyry, which was the usual introduction to logic, and of Aristotle's *Hermeneutica*. These translations cannot be identified, but translations of Aristotle's *Hermeneutica* and *Analytica Priora* as well as of Porphyry's *Isagoge*, with commentary attached exist, made by Probus, who is described as presbyter, archdeacon, and chief physician of Antioch, which seems to be contemporary and it may well be that the version of the text is that of Hibha. 'Abdyeshu' bar Berikha (thirteenth to fourteenth century) speaks of Hibha, Kumi, and Probus as contemporaries and all translators of Aristotle. Of Kumi's version nothing is known. Early in the sixth century, therefore, these works on logic were known at Edessa in Syriac versions. (Syriac vers. of Porphyry, ed. A. van Hoonacker in *J.A.*, xvi, 70–160 ; Aristotle's *Hermeneutica*, ed. G. Hoffmann, Leipzig, 1869, 2nd ed. 1878 ; *Analytica*, ed. J. Friedmann (Erlanger Dissert.), Berlin, 1898.)

(3) The Nestorian Schism

In 428 Nestorius,[3] a monk of Antioch, was made Patriarch of Constantinople, an outsider chosen to avoid inflaming the strong faction spirit prevailing in the capital, which would have been the inevitable result of appointing a local candidate. Nestorius brought with him a brother monk of Antioch Anastasius. Both of these were products of the school of Antioch, trained in the theology of Theodore and Diodorus.

[3] See p. 183.

Before long a sermon preached by Anastasius was made the subject of a complaint to the Patriarch. The objection laid was that Anastasius denied the applicability of the term *Theotokos* to the Blessed Virgin Mary, asserting that she was the mother only of the human body of Christ. To some extent the question was one of psychology : Does the soul enter into man at birth, or is it present before birth?—orthodox fathers have differed in their answer. If the reasonable soul does not enter into the body until birth, it might be assumed that the *Logos*, the Divine Person of Christ, would not have entered his body whilst it was as yet only an animal body, not human until the reasonable soul was added. Anastasius' teaching was not that of Diodorus and Theodore, for they do not seem to have dwelt upon this point. To the populace the refusal of the title Theotokos to the Blessed Virgin seemed blasphemous and passion was inflamed. Beneath this were the rival tendencies to Antioch and Alexandria. Antioch inclined towards what we may call a semi-rationalist treatment of theology, Alexandria towards an allegorical and mystical treatment, and the Alexandrian school had a strong outpost in Constantinople.

When complaint was made to Nestorius he defended Anastasius and the controversy became embittered. As it raged in the capital city, other churches intervened, opposition to Nestorius being stirred up by Cyril the Patriarch of Alexandria. At length the Emperor intervened and a general council was held at Ephesus in 431 at which Nestorius was deprived and excommunicated. But many Syrians disapproved of this decision, repudiated the council, and separated from the orthodox Church. These separatists were known as Nestorians.

The Christian school at Edessa, trained in the theology of Diodorus and Theodore, generally supported Nestorius, although there was a strong minority opposed to his teaching. It became the focus of Nestorianism and in this had Hibha as leader. At first the bishop Rabbula took the Nestorian position, but he was won over by Cyril's arguments and stood out against the teaching prevalent in the school. At his death in 435 Hibha, the head of the school and a prominent Nestorian, was appointed bishop and the policy of Rabbula was reversed.

In the controversy raised about Nestorius his leading

opponent was Cyril of Alexandria, whose opposition admittedly was conducted in a somewhat intemperate manner.   Even at the Council of Ephesus his action was arbitrary, for he pressed the council to commence without waiting for the arrival of the Asiatic bishops, some of whom would probably have supported Nestorius.   When those Asiatic bishops arrived they found that matters were already decided and Nestorius condemned.   Greatly indignant at this having been done in their absence, they held a rival council under the presidency of John, Patriarch of Antioch, and there decreed the deposition of Cyril of Alexandria, and his chief supporter Memnon, Bishop of Ephesus.   The decrees of both councils required the endorsement of the Emperor Theodosius.   He, offended at Cyril's arrogant behaviour, ratified the deposition of Nestorius, Cyril, and Memnon, then changed his mind and permitted Cyril and Memnon to retain their sees, but compelled Nestorius to return to his monastery near Antioch, where he remained until 435, when he was banished to Petra in Arabia, though he seems actually to have been allowed to go to an oasis in Upper Egypt.   Whilst there he was carried off by a nomadic tribe, but escaped and was driven from one place to another by imperial officials, until he died in circumstances unknown some time after 439.

Cyril of Alexandria died in 444 and was succeeded by Dioscoros who followed Cyril's teaching, but surpassed him in violence and autocratic self-assertion.   He at once began to search out and persecute all who could be suspected of any tendency towards Nestorianism.   Then a new dispute was raised by Eutyches, the aged archimandrite of a monastery at Constantinople, who propounded the doctrine that in the Incarnation the humanity of Christ was completely merged in his Deity, and the Nestorians (wrongly) asserted that their opponents were Eutychians.   Eutyches had been a supporter of Cyril, but his teaching was opposed by Eusebius, Bishop of Dorylaeum, who had also been one of Cyril's supporters, and the matter was brought before Flavian, Patrich of Constantinople, and his local synod.   Flavian was one of the Antiochene school, but of the moderate wing, and was drawn into the controversy reluctantly, but at length Eutyches was deposed and excommunicated.   To Dioscoros, who appears to have inclined towards Eutyches' view, or at any rate considered it nearer the truth than the doctrine of Nestorius, this seemed like a revival of Nestorianism and, by the favour of the Empress, he obtained a re-hearing of the case before another synod at Constantinople a year later, but this synod did not reverse the sentence against Eutyches. Dissatisfied with this Dioscoros induced the Emperor to summon a general council for the extirpation of Nestorianism in 449, and at this council he himself presided. But when the council met his conduct was violent and arrogant so that the assembly became a scene of confusion, well deserving the name of *Latrocinium* or " Synod of Brigands ", which Pope Leo applied to it. Eutyches was restored, his accuser, Eusebius of Dorylaeum, was not even granted a hearing, and Flavian was deposed. When some of the bishops present ventured to remonstrate, Dioscoros called in a band of soldiers and threatened them into submission. At this council Hibha, Bishop of Edessa, was deposed and a pronounced Cyrillian Nonnus was appointed in his place.

But the proceedings of the " Synod of Brigands " aroused general disapproval, and those who disapproved most turned to Rome for support. After a great deal of heated controversy another general council was assembled at Chalcedon in 451, and this council, strongly prejudiced against Dioscoros, reversed the decisions of 449, deposed Dioscoros, and drew up a statement of faith which seemed a reasonable compromise. Dioscoros and his partisans refused that statement and separated from the State Church. Thus the Eastern Church was divided into three bodies, the Orthodox or State Church, the Nestorians, and the extreme anti-Nestorians who rejected the Confession of Faith proposed at Chalcedon and are commonly known as Monophysites.

There has been a good deal of antagonism to Hibha's appointment to the see of Edessa and the objectors made their complaint to Domnus who became Patriarch of Antioch in 442. Domnus seems to have been unwilling to listen to this complaint, but in 448 a formal charge was laid in such a form that it could not be ignored and Hibha was summoned to Antioch to answer the accusations brought against him. The synod was held at Antioch after Easter and only a few bishops attended, the extant decrees are signed by nine bishops only. Eighteen charges were laid against Hibha : one of these was that he had anathematized Cyril of Alexandria as a heretic, and this he admitted. Other charges, that he was a Nestorian, that he had uttered blasphemous words in a sermon on Easter Day, 445, and other matters he denied. Of four witnesses who appeared against him two went away to Constantinople because they considered that Domnus was biassed in Hibha's favour, and in their absence the trial was postponed indefinitely. The two who had gone to the capital appealed to the Emperor and the case was remitted to a special commission which was called to meet at Tyre, but this was afterwards changed to Berytus (Beirut). The commissioners declined to come to any definite conclusion and a compromise was effected on 25th February by which Hibha agreed to pronounce an anathema publicly upon Nestorius and to accept the decrees of Ephesus. Such a truce could not be permanent : Hibha's enemies were active and had friends at court, so another council was arranged at Ephesus later in the same year, which was the notorious *Latrocinium*, and at this he was deposed and excommunicated. But the scandal caused by that council brought a change of feeling generally and the Council of Chalcedon in 451 restored him as unlawfully deprived, but required him to anathematize both Nestorius and Eutyches, which he did, and resumed his see. Apparently Hibha's personal character told greatly in his favour and he retained undisturbed possession of his see until his death on 28th October, 457, when Nonnus, who had been put aside at Hibha's restoration, resumed the episcopal office.

When Hibha was appointed bishop he placed his pupil Barsauma, a native of Northern Mesopotamia, in charge of the school. Barsauma shared Hibha's deprivation in 449, and presumably was restored when the Council of Chalcedon revoked the proceedings of the *Latrocinium*. When Hibha died he was still head of the school, and as the leading supporter of Nestorianism was the chief target of Nonnus' persecuting zeal. This became so intolerable that Barsauma decided to leave Edessa and seek a new life in the kingdom of Persia. Whether he was actually expelled is not clear : the opponents of Nestorianism in the school of Edessa were in the minority, but they were a strong minority and now they had the bishop's support. The view has been proposed that the school at Edessa was Nestorian, the city anti-Nestorian.

The history of this period contains several difficulties in its chronology, which are not easily solved. Certain fixed points can be determined from outside sources, and these are : In 435 Hibha became Bishop of Edessa, and apparently entrusted the school there to Barsauma then, or soon afterwards. 449, the *Latrocinium* or " Brigands' Synod " deposed both from office. In that year there was a popular outbreak against Barsauma demanding his expulsion from the city. He was a leading and very contentious Nestorian. There was a strong anti-Nestorian minority at Edessa : it has been suggested that the school was Nestorian, the people generally were not, but this is dubious.

451, Hibha was restored to office by the Council of Chalcedon, probably Barsauma was restored at the same time.

457, Hibha died and his successor Nonnus enforced the Chalcedonian decrees, dealing harshly with the Nestorians. As a result some of the Nestorian lecturers (including Barsauma ?) migrated to Persia.

471, Cyrus became Bishop of Edessa and continued a strongly anti-Nestorian policy.

482, the Emperor Zeno endeavoured to win back the Monophysites who had separated from the Church and published the Henoticon as a compromise. This Henoticon was primarily addressed to the Church of Egypt and in it the Emperor condemned Nestorius, approved Cyril of Alexandria, and neither approved nor rejected the canons of Chalcedon. The imperial government was anxious to conciliate the Monophysites, but do not seem to have troubled much about the comparatively unimportant Nestorians. The Nestorians regarded this as a direct attack upon their religion and were greatly disturbed at the way in which, as they viewed it, the government had gone over to their enemies, the Monophysites.

489, the Emperor Zeno was persuaded by Cyrus, the Bishop of Edessa, to close the school there finally and the lecturers who were Nestorians forthwith migrated to Persia. They were met by Barsauma and induced to settle at Nisibis, where they opened a school entirely Nestorian in its teaching, and this school was directly descended from the school of Nisibis and afterwards became the great central university of the Nestorian community.

There were two definite purges of the school of Edessa,

E

one in 457, the other in 487, all the remaining Nestorians going away after this latter.

The contemporary Persian kings were—

438–457	Yazdgerd.
457–484	Peroz.
484–488	Balash.
488–531	Qawad I.

The contemporary Catholici or metropolitans were—

415–420	Yahbalaha.
420	Ma'na, Farbokht.
421–456	Dadisho'.
457–484	Babowai.
485–495–6	Aqaq (Acacius).
497–502–3	Babai.

The historian Shem'on, of Beth Arsham, says that Barsauma, Aqaq, Ma'na, John, Paul, son of Qaqai, Pusai, Abraham, and Narsai, all lecturers of Edessa, migrated to Persia after Hibha's death (457), were received by Babowai, and settled in Persian sees. Barsauma then set himself to rally the Nestorians and force Nestorianism on the Persian Church. Shem'on is a strongly prejudiced Monophysite.

It seems clear that Barsauma was befriended by Babowai, who presented him to King Peroz, and as the Catholicos vouched for his capacity to negotiate with the Romans, Peroz gave him the oversight of the frontier defences, and subsequently employed him on a commission to check the boundary with the Persian Marzban, the Roman *dux*, and the king of the Arabs. All this must have taken place before the summer of 484 when King Peroz died, and probably before the April of that year when Babowai was executed.

During the period 457–484 Barsauma took drastic measures to promote Nestorianism in Persia. He persuaded the king that it was necessary that the Persian Church should be differentiated from the orthodox Church in the Roman Empire, and one measure he took to do this was to induce the bishops to marry, which fitted in very well with the Persian idea that it was every man's duty to be married and rear children. To enforce this he held a council at Bait Lapat (Jundi-Shapur) in April, 484, a synod attended by only a few bishops, and there decreed the legality of episcopal marriage. The synod was afterwards adjudged to be null and void as Barsauma was not the metropolitan, who alone was entitled to convoke synods, and consequently its decrees are not included in the *Synod. Orient.* No doubt Barsauma counted on being made Catholicos at Babowai's death, but as his protector Peroz died soon afterwards, before the bishops met to elect a new metropolitan, they were able to hold a free election and, already aware that Barsauma was a man of turbulent and tyrannical tempers preferred to choose Aqaq (Acacius), who was also an alumnus of the school of Edessa. The new Catholicos held a synod at Beth 'Adrai in August, 485, at which the canons of Beit Lapat were confirmed, and a more formal council at Seleucia in February, 486, whose acts have come down to us (*Synod. Orient.*, 299–309), and from these we can gather the general tendency of Barsauma's changes designed to adapt the Nestorian Church to Persian standards. All this seems to have been a reaction against the anti-Nestorian development in the Roman Empire under Zeno. Six letters which passed between Barsauma and the Catholicos Acacius are preserved in *Synod. Orient.*, 532–9, and reveal him as a strong opponent of everything hostile to Nestorianism and a devoted servant of the Persian crown.

Narsai, who may have remained at Edessa until the school was finally closed in 489 and have succeeded Barsauma as head of that school, or may have accompanied Barsauma in his migration to Persia before that, as Shem'on of Beth Arsham says, was equally vigorous in his advocacy of Nestorianism, but for a period was opposed to Barsauma and harshly treated by him : undoubtedly Barsauma was a man of overbearing and arbitrary temper. After he was made Bishop of Nisibis (485), probably after the closing of the school of Edessa (489), Barsauma established the school of Nisibis and placed it under the direction of Narsai (cf. below).

Shem'on associates a third person with Barsauma and Narsai as spreading Nestorianism in Persia after 457, a rather obscure character called Ma'na who is described as having ultimately become catholicus. But the only catholicos of this name which appears in the list of the Persian metropolitans was made catholicos in 420, in the last year of King Yazdgerd I, thirty-seven years before the death of Hibha. Shem'on further

describes him as having translated Syriac books into Old
Persian and as making a Syriac translation of the commentary
of Theodore of Mopseustia for Hibha. According to the
Nestorian chronicles Yazdgerd I became a persecutor in the
last year of his reign, urged on by the native priesthood who
were alarmed at the spread of Christianity, which probably
means that many Mazdeans had been converted to Christianity,
contrary to Persian law. So Yazdgerd deposed Ma'na, forbade
him to control the affairs of the church, and relegated him to
his native province. Mare and Elias of Nisibis refer to him
as being banished and imprisoned, but liberated on the
undertaking that neither he nor any other should claim the
title of catholicos. His name does not occur at all in the
diptychs of the Nestorian Church, and the chronicles give
Ma'na, Farbokht, and Dadisho' as becoming catholicos in
420 or 421, but agree that Dadisho' held that office from 421
to 456 and was then followed by Barsauma's friend Babowai.
The most probable solution seems to be that at the death of
the catholicos Yabalaha in 420 there was a disputed election
with three candidates, that Ma'na and Farbokht held their
own for a while, then in 421 Dadisho' obtained general recognition, the comparatively obscure Majna being afterwards
confused with a namesake who left Edessa with Barsauma.

There is another obscure name which sometimes seems to
replace that of Ma'na, Mari the Persian. He, like Ma'na,
is described as of Beit Ardashir, which is the official name for
Seleucia, so it is implied that he was Bishop of Seleucia and
consequently Catholicos. But no catholicos of that name
occurs in the lists of metropolitans. He is said to have corresponded with Hibha, but the catholicos in Hibha's days was
Dadisho'. It has been suggested that Mari stands for Dadisho' :
the term means " lord ", a complimentary title usually prefixed
to the name of the catholicos, which has been accidentally
taken for his name. Admittedly the name Dadisho' was
difficult to transliterate in Greek (cf. Labourt, *Le Christianisme
dans l'Empire Perse*, p. 133, note 6).

The other alumni who migrated from Edessa to Persia are
easier to enumerate. They are, Aqaq (Acacius), who became
catholicos in 485 ; Aba Yazadid ; Yuhanna (John of Beth
Garmai, east of the Tigris), who was made Bishop of Beth
Sari) ; Abraham the Mede ; Paul, the son of Qaqi, who

became Bishop of Beth Huzaye (Ahwaz), and died about 535 ; Micah, who became Bishop of Lashom of Beth Garmai ; Pusi, who became Bishop of Huzaye ; Ezalaya, of the monastery of Kefar Mari ; and Abshota of Nineveh. All these are enumerated with derisive nick-names by Shem'on of Beth Arsham as those who adhered to Nestorian teaching at Edessa after 457, and most of them are described as pupils of Narsai, which may imply that they continued under his instruction after he had removed to Nisibis. All these were Persians, evidently the cream of the theological students of the Persian Church who had been sent to complete their studies at Edessa, the leading Syriac university, and they now returned home : such men probably were marked out for high office in any case.

All this shows the steady transfer of Greek scholarship, in a modified Syriac form, from Edessa across the Persian frontier to Nisibis, whence it ultimately spread through the Nestorian community, and so reached the Arabs. It is a distinct link in the chain of transmission, but a link which at one time almost broke through, and then was renewed. That has now to be considered.

The Greek scholarship transmitted from the school of Edessa to the Persian school of Nisibis consisted mainly of the logical works of Aristotle with the *Isagoge* of Porphyry. The study of the Aristotelian logic was introduced amongst the Syriac-speaking Christians by Hibha, who translated, or procured the translation, of Aristotle's *Hermeneutica* and *Analytica Priora*, with Porphyry's *Isagoge*, and these were soon circulated with the commentaries of Probus (*c.* 450), independent of the Greek commentators but with some use of Ammonius. At a later date the Nestorians employed Ammonius' commentary, whilst the Monophysites preferred that of John Philoponus. In the first place Hibha had introduced the Aristotelian logic to illustrate and explain the theological teaching of Theodore, of Mopseustia, and that logic remained permanently the necessary introduction to theological study in all Nestorian education. Ultimately it was that Aristotelian logic which, with the Greek medical, astronomical, and mathematical writers, was passed on to the Arabs.

Barsauma is stated to have composed metrical homilies, hymns, and a liturgy. His most interesting literary production

is the series of six letters which he wrote to the Catholicos
Acacius, fortunately preserved in the *Synodicon Orientale*,
which has been edited by J. Chabot, with transl. and notes,
(Paris, 1902.)

Narsai, whom Barsauma placed in charge of the restored
school of Nisibis, was a voluminous writer, though only
fragments of his works survive. 'Abdisho' ascribes to him
scriptural commentaries, 360 metrical homilies, a liturgy,
expositions of the Eucharistic liturgy and Baptism, and various
hymns of which two are often included in the Nestorian
Psalter (Daily Office).

Narsai died probably between 500 and 520, and was
succeeded by his nephew Abraham. Of his pupils the best
known were John of Nisibis and Joseph Huzaya, who died
about 575. John of Nisibis was the author of a number of
commentaries on scripture and other theological works :
" If the discourse on the plague at Nisibis and the death of
Khosraw I. Anoshirwan be really by him, he was alive in 579
in the spring of which year that monarch died " (Wright,
Hist. Syr. Lit., 115). Joseph Huzaya was the first Syriac
grammarian (cf. Merx, *Hist. artis grammat. apud Syros*, Leipzig,
1889, pp. 26 *sqq.*).

(4) Dark Period of the Nestorian Church

In passing through the medium of a foreign language any
form of intellectual culture is liable to suffer modification,
though this may be merely superficial, and such undoubtedly
was true of Greek scholarship as it passed through Syriac
translations. But this change was most pronounced in the
Nestorian atmosphere, for that became more definitely oriental
after Barsauma's deliberate policy of Persianizing the Nestorian
Church. His efforts resulted in making a great cleavage between
Greek Christianity as it existed within the Roman Empire,
and Nestorian Christianity at home in Persia. The Nestorian
schism had already made a division in doctrine : the synods
of 484 and the following years made a great difference in
discipline until they were repealed in 544 : in worship a
divergence arose from the fact that the Nestorians after 457
were out of touch with the liturgical life of the Eastern Church
at large, and this was accentuated by the compilation of special

liturgies by Barsauma and others : politically there was a cleavage because the Greek Church remained under the imperial government at Byzantium, whilst the Nestorians were subjects of the Persian King : and culturally a separation arose from the fact that students, theological or other, ceased to visit for study those lands where Greek was still a living language. This cleavage, begun by Barsauma, became wider under his immediate successors.

Acacius and his successor Babai had received an education which, though Syriac in form, was Greek in substance. After that the episcopate rapidly became more Persian, and as it orientalized it degenerated.

The discipline of the Eastern Church encourages a married secular (parochial) clergy, married before ordination, marriage after ordination and second marriages not being permitted : monks and nuns are of course celibate, and bishops and certain other dignitaries are chosen only from the (unmarried) regular clergy. Hormizd III, son of Yazdgerd II, who for a brief period occupied the Persian throne after his father's death and was then replaced by Peroz, had persuaded the Catholicos Babowai to marry a girl of great beauty whom he selected, holding the Persian opinion that it was every man's duty to marry. Babowai could not refuse, but at once sent back the damsel to her family. Peroz, in his friendship for Barsauma, acted similarly : Barsauma could not refuse, but kept his bride though abstaining from marital relations with her, according to the Nestorian historians. But Barsauma, desiring to differentiate the Nestorians from the Greeks and wishing to please the king, advised that the bishops be permitted to marry, even after ordination : he desired that Christian clergy should enjoy a good repute in the eyes of the pagans and their magi.

Barsauma's policy resulted in the canons passed by the council held at Seleucia in 486. After affirming Nestorian doctrine (canon 1), it was decreed that monks may not intrude in towns where there already are parochial clergy or minister the sacraments, they must remain in their monasteries or desert hermitages (canon 2), the vow of celibacy binds only cloistered religious and no other clergy, those already deacons may marry, and no more persons may be ordained deacons unless they are married and have children, and priests like all other Christians are allowed to contract second marriages.

From 486 until those canons were repealed, the Persian (Nestorian) Church was undoubtedly orientalized and was regarded by the rest of Christendom as a degenerate by-product of Christianity.

The death of Barsauma did not check the Persianization of the Nestorian Church, and a council held at Seleucia in 499 formally approved the marriage of the Catholicos, the bishops, and priests.

At the death of the Catholicos Babai in 502 or 503 there followed a period of anarchy when the Persian bishops were unable to agree on the appointment of a metropolitan. At last Babai's archdeacon Shila was appointed, chiefly because he was a favourite of King Qawad. But he did not turn out well, he disposed of church property to his son and designated his son-in-law Elisha as his successor, a kind of nepotism likely enough to arise among married clergy. At Shila's death in 523 a number of bishops elected Narsai, the Bishop of Hira, as Catholicos and consecrated him at Seleucia. But Elisha had his partisans and they held a rival consecration at Ctesiphon close by Seleucia. Thus the Nestorian Church was split, each section appointing its own bishops and clergy and excommunicating the opposing party. About 535 Narsai died, but his partisans elected and consecrated Paul, the archdeacon of Seleucia, and so the schism was continued. Paul, however, was an old man and died two months after his consecration, and then the Narsai party elected Maraba, who was destined to be the reformer of the Nestorian Church and the leader of a revival of learning which would restore the scholarship of Edessa. It is worth while sketching this history, petty as some of its details may appear, as it shows how far the Nestorian community had degenerated and disintegrated under Persian rule, entirely cut off from intercourse with the main stream of Christian life and Greek scholarship.

(5) THE NESTORIAN REFORMATION

Maraba was a native of the country west of the Tigris. As to religion, he had been brought up in the Mazdean faith and after holding the office of *arzbed* of his town under the Persian government, had been promoted to the post of assistant secretary to the *hamaragerd* of Beth Aramaye. There he met a Christian catechist named Joseph, who had been a pupil in the school of Nisibis and, as they travelled together he treated him with disdain because he was a Christian, but was overcome by his humility and readiness to help when they were in a difficult position at a flooded river. After that they began conversing and discussed matters relating to their respective religions with the result that Maraba was baptized a Christian. Then he went to the school at Nisibis and attached himself to a teacher named Ma'na. When Ma'na was made Bishop of Arzun, Maraba went with him to his see and was active in preaching to pagans and heretics. After this he returned to Nisibis and completed his studies there. Then he set out to travel in the Roman Empire so as to obtain a better knowledge of the Greek language in which so much material relating to the Christian religion was written. At Edessa he met a Syrian named Thomas who gave him instruction in Greek, and together the two visited the holy sites in Palestine and the hardly less holy sanctuaries of Shiet (Scetis) in Egypt, the cradle of the monastic life. Finally he returned to Persia, but was so shocked at the state of the Nestorian Church and the schism which divided it, that he prepared to devote himself to a hermit's life like that of the ascetes whom he had seen in Egypt. But the bishops intervened and forbade him, insisting that he should undertake teaching, then after a while they elected him Catholicos, exhorting him to counteract the threatened encroachment of Monophysite propaganda. His first task was to restore discipline in the church, then he turned to the promotion of scholarship and especially to the study of Aristotelian logic. To further this he founded a school at Seleucia, for there seems no basis for a legend which claims an earlier foundation for that school, and this school of Seleucia had a reputable history, but it never became a serious rival to the older school at Nisibis which remained the central university of Nestorian Christianity.

Maraba's episcopate lasted from 536 to 552. Unfortunately his great activity aroused jealousy and he had a quarrel with King Khusraw I with the result that the king had the Nestorian church at Seleucia pulled down and sent Maraba into exile to Adharbaigan (Azerbaijan). As Maraba was a convert from the Mazdean religion he was of course liable to death, but he was by no means the only such convert who escaped that penalty.   He returned from exile without permission, was cast into prison, and died there on 29th February, 552. His body was removed to Hira [4] and buried there, and a monastery was erected over his grave.   This Arab city of Hira was by now a great stronghold of Nestorianism.   He is said to have attempted a revision of the Peshitta or Syriac vulgate of the Old Testament, perhaps of the New Testament as well, but the Nestorians generally clung to the older version to which they were accustomed.   He was the author of commentaries on Genesis, Psalms, Proverbs, and the epistles of St. Paul, of homilies, hymns, synodal epistles and canons, these last strongly against the marriage of bishops and priests. His influence generally was the revival of life in the Nestorian Church and a return from oriental isolation to a closer contact with the Greek Church.

In his days there lived two writers, both known as Abraham of Kashkar.   One of these was a student of philosophy and also a reformer of monasteries.   He is said to have written a treatise on the monastic life which was translated into Persian by his disciple Job the Monk.   His namesake was a student of Nisibis, and he also was a monastic reformer.   He preached in Hira and converted many pagan Arabs, then went to Egypt and Sinai, finishing his life as a hermit on Mount Izla.   He left a code of monastic rules considerably stricter than those previously accepted in the Nestorian monasteries.

Theodore of Marw was appointed Bishop of Marw by Maraba in 540.   He was a disciple of Sergius of Rashayn who is reckoned as a Monophysite (cf. *infra*), and like his teacher was a student of Aristotelian logic.   In him and the first Abraham of Kashkar we have evidence of the humanist renaissance which was taking place in Maraba's days, amongst Monophysites and others as well as in Nestorian circles, but to which he was the chief agent in directing the Nestorians. Theodore's brother Gabriel was Bishop of Mormuzd-Ardasher (Ahwaz), and has also left literary works, but those were entirely theological, commentaries on scripture, and a treatise against the Manichaeans and the astrologers.

With the revival of the school of Nisibis the Nestorians started a system of general education in schools attached to their churches, and in these children were taught hymns and

[4] Cf. page 184.

church music. The school of Nisibis itself was in the form of a coenobium where the students were bound by vows of celibacy, continuous residence, regularity, and diligence. Not all were monks or intending to be monks, and these vows and monastic discipline bound them only so long as they attended the school. The head of the school for some time after Narsai's death was *Henana* of Adiabene and under him, it is said, there were 800 students in attendance. But early in the seventh century the school was vexed by dissentions caused by those who wanted reform, restoration of stricter discipline, and the more definite Nestorianism which had prevailed under Barsauma, for Henana had taught a modified form of Nestorian doctrine which compromised with the teaching of the Orthodox Church. His teaching had a considerable following, but was opposed by many, so the Persian Church generally was divided and this division was reflected in the school. Some of the dissatisfied left Nisibis and founded other schools more in conformity with their ideas in the monasteries of Abraham and Bath 'Abe, but these never became serious rivals of Nisibis. Under the Catholicos Isho'yahb (628–643) the desired reforms were introduced into Nisibis and the schism was healed. The school was flourishing at the time of the Muslim conquest, but does not seem to have had any direct influence on the Arabs, probably because it was so definitely theological, though it no doubt was indirectly responsible for introducing the logic of Aristotle to the other Nestorian academies at Jundi-Shapur and Seleucia. The influence which reached the Arabs came mainly through Jundi-Shapur.

The rivalry of Monophysite propaganda not only prompted a revival of learning amongst the Nestorians, but also suggested an expansion into the surrounding country where their Monophysite rivals were winning many converts from the pagan Arabs. Thus began the missionary enterprise of the Nestorians which before long spread amongst the Arabs on the south-west, then eastwards across Central Asia until it reached the Far East.

On the Persian border the chief city of the Arabs was Hira. Towards the end of the sixth century Nu'man, King of Hira, was baptized, and this was followed by the conversion of many of the Arabs. In Hira these Arabs, of the Lakhmid clan, formed the ruling aristocracy, the bulk of the population was

Aramaic Syriac and already Christian. It appears that those Arabs who accepted Christianity embraced Nestorian doctrine, accepted the ministrations of Syriac-speaking Nestorian clergy, and used Syriac as a liturgical language. As yet there were no books in Arabic, no Arabic version of the scriptures, and no Arabic liturgy. It appears that Hunayn ibn Ishaq, who was a native of Hira, had to learn Arabic later in life, the humbler classes of Hira being Syriac-speaking.

Nestorian missions pushed on towards the south and reached the Wadi l-Qura', a little to the north-east of Medina, an outpost of the Romans garrisoned, not by Roman troops, but by auxiliaries of the Qoda' tribes. In the time of Muhammad most of those tribes were Christian, and over the whole wadi were scattered monasteries, cells, and hermitages. From this as their headquarters Nestorian monks wandered through Arabia, visiting the great fairs and preaching to such as were willing to listen to them. Tradition relates that the Prophet as a young man went to Syria and near Bostra was recognized as one predestined to be a prophet by a monk named Nestor (Ibn Sa'd, *Itqan*, ii, p. 367). Perhaps this may refer to some contact with a Nestorian monk. The chief Christian stronghold in Arabia was the city of Najran, but that was mainly Monophysite. What was called its Ka'ba seems to have been a Christian cathedral.

But Greek culture did not pass through these early contacts. The cultural contribution of the Nestorians was definitely through Jundi-Shapur, and the transmission of Greek science to the Arabs took place when the Arab court was established at the newly built city of Baghdad close by.

The pontificate of Maraba fell within the reign of Khusraw I (531–578). Although that king conducted a war against the Romans, he was a great admirer of Graeco-Roman culture and especially desired to introduce Greek science into his dominions. It was he who offered hospitality to the philosophers who were turned adrift when Justinian closed the schools of Athens and provided for their safety and welfare when they decided to return to Greece. He desired to have in Persia a great Greek academy like that at Alexandria, and such an academy he established in the city of Jundi-Shapur. There the Alexandrian curriculum was introduced and the same books of Galen read and lectured upon as at Alexandria. This was no new departure, for the same curriculum was followed at Emesa where there also was a school. Obviously the courses followed at Alexandria were in great repute and were generally regarded as the model for a secular education.

Greek physicians greatly esteemed certain herbs and drugs which could only be obtained from India, and so Khusraw sent an agent, *Budh*, a Christian periodeutes or rural bishop, to India to procure drugs. To this Budh is ascribed a work which was called *Alef Migin*, which has been explained as meaning a commentary on the first book of Aristotle's *Physica* (Ἄλφα τὸ μέγα), which is not extant, and the Syriac version of a collection of Indian (Buddhist) tales known as *Qalilag wa-Dimnag*, but " that Bodh made his Syriac translation from an Indian (Sanskrit) original, as 'Abdh-isho' asserts, is wholly unlikely ; he no doubt had before him a Pahlavi or Persian version " (Wright, *Hist. Syr. Lit.*, 124). It is also stated that Khusraw brought a physician from India to teach medicine in the Indian fashion and established him at Susa, meaning of course Jundi-Shapur. Nothing is known of that physician, neither his name nor any details of his activities. To judge by the appendix on Indian medicine attached to the " Paradise of Wisdom " (*Firdaws al-Hikhma*) of 'Ali b. Sahl b. Rabban at-Tabari (*circ.* 850), Indian medicine at that time did not amount to much, it was largely concerned with the exorcism of evil spirits supposed to be the cause of disease with some theories of a confused and vague psychology (cf. *Firdaws al-Hikhma*, ed. W. Z. Siddiqi, Berlin, 1928). It is possible that for Khusraw I Persian translations were made of portions of Aristotle and the *Timaeus*, *Phaedo*, and *Gorgias* of Plato. Agathias heard of such translations, but did not believe in their existence.

Under Khusraw I lived *Paul the Persian* (d. 571) who " is said by Bar Hebraeus to have been distinguished alike in ecclesiastical and philosophical lore and to have aspired to the post of metropolitan bishop of Persia, but being disappointed to have gone over to the Zoroastrian religion. This may or may not be true. . . . Bar Hebraeus speaks of Paul's " admirable introduction to the dialectics (of Aristotle) ", by which he no doubt means the treatise on logic extant in a single MS. in the Brit. Mus. (Add. 14660, f. 55*b*) (Wright, *Hist. Syr. Lit.*, 122–3). This is edited in Land, *Anecd. Syriaca*, iv, text 1, 32, trans. 1–30).

There was a Persian academy at Raishahar in the Arrajana
province where work was carried on in medicine, astronomy,
and logic, which suggests another reproduction of the
Alexandrian curriculum (Yaqut, *Muajjan ul-buldan*, ed.
Wüstenfeld, ii, 887, trans. Barbier de Maynard, *Geographical,
Historical and Literary Dictionary of Persia*, 270–1). Mention is
also made of an academy with an extensive library at Shiz,
also in Arrajana (Ibn Hawqal, ii, 189, 1–2). But very little
is known of these Persian academies or of Persian physicians
of pre-Islamic times save the names given in a scanty catalogue
by Mansur Mowafih who lived in the earlier part of the tenth
century.

Syriac study of Aristotle was limited to the logic and with it
were taken the *Isagoge* of Porphyry and a compendium of
Aristotelian philosophy by Nicolaus of Damascus, who was
also the author of a Botany which was for some time accepted
by the Arabic students as genuinely Aristotelian. The logic
was read with the help of a commentary, at first the Syriac
Probus (cf. above), later the Greek commentary of Ammonius
or that of John Philoponus, the Nestorians preferring the
former, the Monophysites using the latter. In these commentaries a neo-Platonic influence is already apparent, and
that influence passed through the Syriac versions and commentaries to the Arabs.

From the time of Maraba onwards there is fairly continuous
evidence of translation from the Greek and of work in
Aristotelian logic. Restricting ourselves for the moment to
Nestorian writers we may note :—

Maraba II (more usually *Aba* simply, as he preferred to
distinguish himself from his greater namesake), Catholicos
from 741 to 751, often called Aba of Kashkar, as he was bishop
of that city before he was appointed Catholicos. He is said
to have been skilled in philosophy, medicine, and astronomy,
which sounds like the full Alexandrian curriculum, and to
have been learned in the wisdom of the Persians, Greeks, and
Hebrews (A. Scher, *Chron. de Séert, P.O. VII*). He is
credited with having written a commentary on the Dialectics
of Aristotle. As Catholicos he had a dispute with his clergy
about the management of the school of Seleucia and in this
seems to have fared ill, as he left the city and resided elsewhere
for some years, but finally returned. 'Iraq was conquered

by the Arabs in 638, and Persia in 642. During the whole of the episcopate of Maraba II, Mesopotamia and Persia were under the rule of the 'Umayyad khalifs of Damascus, so it is obvious that the Arab conquest did not check or interfere with the progress of Aristotelian studies which continued in the Nestorian Church under Muslim rule.

Shem'on of Beth Garmai, in the early seventh century is said to have translated Eusebius' Ecclesiastical History into Syriac, but his work is lost.

Henan-isho' II, Catholicos from 686 to 701, is said to have composed a commentary on Aristotle's *Analytica*.

Reference has already been made to Khusraw I's efforts to procure Indian drugs. Amongst those brought to Jundi-Shapur from India was *sukkar* (Pers. *shakar* or *shakkar*, Sanskrit *sarkara*), our sugar, unknown to Herodotus and Ktesias, but known to Nearchus and Onesicritus as "reed honey", supposed to be made from reeds by bees, the μέλι καλάμινον of Theophrastus. Legend relates that Khusraw discovered a store of sugar amongst the treasures taken in 627 at the capture of Dastigird. The juice of the sugar cane was purified and made into sugar in India about A.D. 300, and now the cane began to be cultivated about Jundi-Shapur, where there were sugar mills at an early date. At that time and for long afterwards sugar was used only as medicine, it was not until much later that it began to replace honey as an ordinary means of sweetening. In addition to the medical faculty which had a hospital attached, Jundi-Shapur had also a faculty of astronomy with an observatory, which again follows the Alexandrian model. The study of mathematics was subsidiary to astronomy.

At the time of its foundation as a prisoners' camp Jundi-Shapur had citizens who spoke Greek, others who spoke Syriac, and there must have been some using Persian, as it was so close to the royal city of Susa. But in course of time Greek seems to have been abandoned and academic instruction was given in Syriac, as at Nisibis and other Nestorian academies, which does not necessarily imply that the study of Greek was abandoned. The needs of the teaching staff led to the preparation of Syriac translations of the set books of Galen, portions of Hippocrates, some of the logical treatises of Aristotle, the *Isagoge*, and probably some astronomical and mathematical works, translations made during the period between the days of Hibha at Edessa and Hunayn ibn Ishaq at Baghdad. Hunayn speaks of these translations as bad, but that need mean no more than that they fell very below the standard of his own work.

Ibn Hawqal (*Bib. Geogr. Arab.*, ii, 109–110) says that the people of Jundi-Shapur used the speech of Khuzistan which was neither Hebrew, Syriac, or Persian, and the *Maahiju 'l-fikar* refers to the people there having a jargon (*ratana*) of their own. This must refer to the colloquial of the street, not to the language used in the classroom where Syriac was in use, as is obvious from the fact that Syriac translations were made for the use of lecturers.

When Baghdad was founded in 762 the khalif and his court became near neighbours of Jundi-Shapur, and before long court appointments with generous emoluments began to draw Nestorian physicians and teachers from the academy, and in this Harun ar-Rashid's minister Ja'far ibn Barmak was a leading agent, doing all in his power to introduce Greek science amongst the subjects of the khalif, Arabs, and Persians. His strongly pro-Greek attitude seems to have been derived from Marw, where his family had settled after removing from Balkh, and in his efforts he was ably assisted by Jibra'il of the Bukhtyishu' family and his successors from Jundi-Shapur. Thus the Nestorian heritage of Greek scholarship passed from Edessa and Nisibis, through Jundi-Shapur, to Baghdad.

CHAPTER VI

THE MONOPHYSITES

(1) BEGINNINGS OF MONOPHYSITISM

THE decisions of Ephesus and the excommunication of Nestorius and his adherents did not bring peace to the Church. Before long new troubles arose. It is necessary to follow these at least in outline as they led to further schism in the Eastern Church, and it was the schismatic bodies which separated from the Church which were the means of transmitting Greek learning to the Arabs. When at length the Muslim Arabs invaded the Roman Empire, these sectarian bodies welcomed them as deliverers and were on friendly terms with them. The position is not fairly viewed if we class Christians on one side, Muslims on the other, without further qualification. For some centuries before the Arab invasions the Christians were split into hostile sects, active in spreading their propaganda against one another, in close contact with the Arabs, and so far as the two dissenting sects were concerned, both actively persecuted by the Byzantine government and both in consequence disloyal to it. It is necessary to appreciate this position to understand the relation between the Arabs and the Christians.

In 444 Cyril of Alexandria, the great opponent of Nestorianism, died and was succeeded by Dioscoros, a man of precisely similar views, but much more violent in temperament and hasty in expression, an extremer opponent of Nestorianism lacking the tact which had been Cyril's saving quality. Not long after his accession to the see of Alexandria trouble began in Constantinople. An aged and greatly respected archimandrite or abbot of a monastery there, full of zeal against Nestorianism, committed himself to a new statement of what he believed to be the orthodox faith, asserting that in Christ there were two natures, but that both were fused in one, the human nature absorbed in the Deity. Complaint was made that this was inaccurate and overstated what Cyril had maintained. It is uncertain who made the complaint in the first place, whether Theodoret, or Eusebius of Dorylaeum,

F

or Domnus of Antioch, but it was one of these, all supporters of Cyril and maintainers of the decrees of Ephesus. Whoever did make the complaint it was one who like Eutyches himself had been Cyril's supporter, so the dispute broke out amongst the anti-Nestorians themselves.

The complaint was made to Flavian, who was then Patriarch of Constantinople, a man of the Antiochene school but of moderate views and very reluctant to be drawn into the controversy. Unwillingly he assembled his local synod in 448 and that synod decided that Eutyches [5] must be deposed and excommunicated. To Dioscoros, who seems to have inclined to Eutyches' opinion, or at any rate considered it nearer the truth than the teaching of Nestorius, this seemed like a revival of Nestorianism, a betrayal of the decisions of Ephesus and, by favour of the Empress, he obtained a re-hearing of the complaint before another synod of Constantinople. Both sides had appealed to popular opinion, and Eutyches had placarded the streets with statements of his case in which he alleged that his accusers had falsified the acts of the late synod of Constantinople so this new council was mainly concerned with that charge and decided that Eutyches had not made it good. Again the decision was against him.

But Dioscoros had influence at court and induced the Emperor Theodosius II to summon a general council for the extirpation of Nestorianism. The summons of this new council was dated 30th May, 449, and the council met at Ephesus in the following August. Dioscoros presided at this council but behaved in a tyrannical and arrogant manner, relying on court support, and introducing military guards to enforce his authority. The assembly became a scene of disorder, well deserving the name of *Latrocinium* or " synod of brigands " which Pope Leo applied to it. Eutyches was restored, his leading accuser Eusebius of Dorylaeum was not even granted a hearing, and Flavian was deposed. Some of the bishops present ventured to remonstrate, but Dioscoros called in a band of soldiers and threatened them into submission. It was at this council that Hibha of Edessa was deposed and a pronounced anti-Nestorian Nonnus appointed in his place.

The proceedings of the " Synod of Brigands " aroused general disapproval, and those who disapproved most turned to

[5] See note p. 185.

Rome for support. A great deal of heated controversy followed until July, 450, when Theodosius died and Pulcheria, the late emperor's sister, raised her husband Marcian to the throne. This entirely changed the attitude of the court on which Dioscorus had relied. Marcian desired peace and was prepared to welcome a working compromise which would put an end to the discord which not only distracted the church but was the source of much disorder in the capital.

To effect a settlement he summoned another general council which met at Chalcedon in September, 451, and passed very carefully worded decrees which steered between the teachings of Nestorius and Eutyches (cf. Labbe, iv, 562, etc.), indeed a most cautious and judicious, but perfectly definite, statement of the traditional faith of the Church. Such a statement might have been expected to reconcile all but the extremists. But it was a failure, for the opposition was incoherent, the opponents were the *acephaloi*, the headless, without leader, for Eutyches was disowned, and without programme. A disunited and disordered group of malcontents, in themselves weak, but very difficult to attack. This was the end of the first phase of what was afterwards called Monophysitism, a scattered rather incoherent opposition to anything which tended towards Nestorianism, but all the opponents divided amongst themselves. The one point on which they did to some extent agree was that the council of Chalcedon had rather inclined towards Nestorianism, and this feeling was strongest in Egypt. The dissidents did agree in disliking the recent council.

(2) THE MONOPHYSITE SCHISM

At the close of the council of Chalcedon, Monophysitism entered on its second phase, still incoherent and divided but agreed in opposing the decrees of Chalcedon, a merely negative and protestant position, therefore weak.

Theodosius, a monk who had attended the council and was very dissatisfied at its decisions, went home to Palestine and published his unfavourable comments, with the result that there were riotous outbreaks and bloodshed in Palestine. Dioscoros refused to recognize the decrees of the council and was accordingly deposed, a " Chalcedonian " named Proterius being placed in his stead. But Proterius could only appear in

public with a guard of soldiers, riots broke out in Alexandria
and he was compelled to leave the city. Clearly it was going
to be no easy task to enforce the Chalcedonian decrees. Egypt
and a great proportion of the monks in all parts were definitely
determined to resist. And yet they had no leader, nor any clear
statement of principles on which they were agreed. The
imperial government tried to bring pressure to bear, but was
disinclined to go too far. Prospects seemed altogether
insecure.

At Marcian's death in 457, a military tribune Leo of Thrace
was elected emperor and proved himself both temperate and
firm. He relaxed Marcian's policy and ceased to apply pressure
on those who resisted the decrees of Chalcedon, so that there
was comparative toleration for the dissidents. Dioscoros had
died in banishment at Gangra in Paphlagonia in 454, and
Proterius had fled from Alexandria, so a new patriarch was
appointed, Timothy Aelurus, a monk who had been banished
for resisting Proterius. He was himself banished in 460, but
for the most part the anti-chalcedonians were unmolested and
used the opportunity to establish themselves firmly.

At Leo's death in 474, the throne passed to his grandson
Zeno, who was even better disposed towards the anti-
Chalcedonians than his predecessor, and entertained hopes of
winning them back to the Church, a policy which might have
been possible if the dissenters had a responsible head with
whom he could negotiate, or a coherent syllabus of what they
wanted. To do this he issued, in 482, a declaration known as
the *Henoticon*, primarily addressed to the Egyptian Church
but applicable to all who protested against Chalcedon. This
document condemned Nestorius, approved Cyril of Alexandria,
and neither approved nor rejected the decrees of Chalcedon.
It was a distinct move in the anti-Chalcedonian direction and
held out terms of agreement for the objectors. No particular
attention was paid to the Nestorians, who by now were regarded
as of no great importance. At once the weakness of the opposition appeared. Some of the opponents were ready to accept
the Henoticon, others objected to it as pro-Nestorian. In 476,
there had been a revolt of Basiliscus, Leo's brother-in-law, but
this had been suppressed and Zeno restored. During his brief
usurpation Basiliscus had received anti-Chalcedonian support,
so that by this time sectarian strife had begun to weigh in the

politics of the empire which, no doubt, disposed Zeno to make terms with the schismatics. Opposition to Chalcedon was gathering force. It was about this time that the Armenian Church cast in its lot with the dissentients. Zeno went as far as possible in the anti-Chalcedonian direction short of declaring himself one of the dissenters. Timothy Aelurus died in 477 and was succeeded by Peter Mogus who accepted the Henoticon, so Alexandria though still anti-Chalcedonian was prepared to accept the *via media*.

Zeno died in 491 and his widow married an elderly courtier named Anastasius who, on this account, was elevated to the imperial throne. He reigned twenty-seven years and steadily followed a cautious policy which aimed at preserving the *status quo*. Egypt, by accepting the Henoticon, was partially pacified, though many there disapproved the terms proposed by Zeno, whilst in Syria there was a strong dissenting element, and from Syria now came the first indication of leadership for the schismatics.

In 512 the see of Antioch was vacant, and a monk named Severus was chosen as patriarch. He had been educated a pagan and in his earlier years had become a lawyer, then was converted to Christianity and immediately joined the anti-Chalcedonian faction. It is often the inclination of converts to go to the extreme, and he was no exception. Before long he became a monk and entered a monastery near Gaza and came in contact with Peter the Iberian, Bishop of Gaza, who had been one of the consecrators of Timothy Aelurus. As a thorough-going anti-Chalcedonian Severus repudiated the Henoticon and refused to recognize Peter Mongus as lawful Patriarch of Alexandria. Then he left Gaza and went into an Egyptian monastery, where exactly is not known, under an abbot named Nephalius, but after some time was expelled from that convent. Why is not clear : was he too extreme in his views ?—or was he a disturber of the peace, as he was afterwards said to be elsewhere ? After his expulsion he went to Alexandria and there was the cause of several riotous incidents. At the head of a band of monks he became prominent in destroying several pagan temples, an illegal proceeding as disused temples were supposed to be under imperial protection. In these proceedings, apparently, most of the monks who accompanied him were able to speak Coptic only, not Greek. Was he also Coptic-speaking ?—if so he must have been very familiar with Egypt and the Egyptians. These proceedings in Alexandria made it expedient for him to flee to Constantinople where again he was associated with outbreaks of disorder. It must be borne in mind that our knowledge of this period of his life is derived almost exclusively from the accounts left by those who were his uncompromising enemies, and the age was one when controversy was very bitter and invective unscrupulous : there was no law of libel and those who have written their accounts of Severus were unsparing in their abuse, much of which must therefore be discounted.

But Constantinople did not prove quite so happy a place as Severus hoped when, in 511, Macedonius, a loyal Chalcedonian, was appointed patriarch. The next year, however, Severus was himself appointed patriarch of Antioch and at once left the metropolis to assume his see. His first act as bishop was to pronounce a public anathema on the decrees of Chalcedon, thus declaring himself one of the extremer schismatics. He then claimed to be in communion with Timothy of Constantinople and John Niciota (of Nikiu) who had become patriarch of Alexandria in 507. In this connection he interchanged synodical letters with Alexandria, and this interchange has been continued to the present day. As metropolitan of Syria his hand was heavy on the " Chalcedonians " and he distinguished himself as a persecutor, but here again our information is derived exclusively from those who were his enemies. During the seven years he occupied the patriarchal throne of Antioch, until the death of the Emperor Anastasius, the anti-Chalcedonian party was in the ascendant, and Severus was generally recognized as its leader and spokesman. But for all that not all of that party were at one with him. For the moment we pause before fortune changed and the dissidents began to suffer persecution.

One of the methods employed to promote views adverse to the decisions of Chalcedon was the circulation of spurious works professing to be the writings of Dionysius the Areopagite, the friend of St. Paul. These works were really produced about 482–500, probably in Egypt, and are strongly tinctured with neo-Platonic theories. Whether the writer was one of the party opposed to Chalcedon, or a writer with sympathies with that party, their bias is obvious. These pseudo-Dionysian writings consist of four treatises entitled " On the heavenly hierarchy ", " On the ecclesiastical hierarchy," " On the names of God," and " On mystic theology ". In addition to these treatises there are ten letters, or fragments of letters, with an eleventh existing only in a Latin version and certainly a forgery of much later date. No reference to such works occurs before the sixth century when they were mentioned by Severus of Antioch and Ephraem, who became patriarch of Antioch in 526. The anti-Chalcedonians appealed to them in a conference with the Catholics in 531, but Hypatius the metropolitan of Ephesus asserted "ostendi non posse ista vera ess quae nullus antiquus memoravit" (Mansi, *Concilia*, viii, 817). Subsequently there were many in the eastern church who expressed doubts as to their authenticity, but Severus and his party generally accepted them. They were translated into Syriac by Sergius of Rashayn (d. 536) and seem to have had a good deal of influence in propagating Severus' teaching in Syria.

Akin to these pseudo-Dionysian documents were certain works ascribed to Hierotheus, the reputed teacher of Dionysius the Areopagite. These were not of Greek origin but original compositions in Syriac by one *Stephen bar Sudhaili* of Edessa, a contemporary of Philoxenus. Like the pseudo-Dionysian writings they were tinged with neo-Platonic ideas and exercised an influence over the sectaries, an influence which they passed on to the Arabs at a later date. Stephen was a monk greatly esteemed for his piety. He made a pilgrimage to Egypt, the home of monasticism, and there came under the influence of some heretical monks, including some who had revived the teachings of Origen. On his return to Syria he began teaching the doctrines he had learned in Egypt and was expelled from his monastery for doing so. He then went to Jerusalem where he continued teaching his peculiar ideas, apparently in association with some Origenist monks already settled there. Following Origen he maintained that the fire of hell is not eternal but merely purgatorial so that ultimately the population of hell will be redeemed and God will be all in all (1 Cor., xv, 28). Theodosius of Antioch (887–96) wrote a commentary on " the Book of Hierotheus " (Brit. Mus. Add., 7189).

We have now reached what we may regard as the close of

the second period of the anti-Chalcedonian movement, the period during which it enjoyed court favour, because it was hoped that the dissenters might still be reconciled with the Church, and during which it showed that it was predominant in Egypt and very strong in Syria. This period ended with the death of the Emperor Anastasius on 11th July, 518.

(3) PERSECUTION OF THE MONOPHYSITES

At the death of Anastasius Justin, a Thracian peasant, installed himself as emperor. The anti-Chalcedonian party in Constantinople was led by the eunuch Amantius, who was determined to set Theocritus on the throne, but entrusted the distribution of largesse to Justin, and Justin made good use of the influence this gave him, so that he was able to secure the throne for himself. This new emperor was a Catholic and orthodox; that is to say he accepted the decrees of Chalcedon and determined to enforce them. A council was held in Constantinople on 20th July, 518, at which it was determined to reverse the policy of Anastasius and Zeno and to enforce conformity with Chalcedon. This new policy was endorsed by a synod at Jerusalem on 6th August, and by another at Tyre on 14th September.

Severus of Antioch was regarded as the leading opponent of Chalcedon and orders were sent for his arrest, but he escaped and took refuge in Egypt. At the same time orders were issued for the deposition of all anti-Chalcedonian bishops, and a number of these, including Julian of Halicarnassus, also took refuge in Egypt. Egypt was too great a stronghold of the opponents to be dealt with, and for the while it was left alone. When Severus arrived there Dioscorus II, who had succeeded John Niciota in 517, was patriarch, but he died on 24th October, 518. Pope Hormisdas advised Justin to take the opportunity of restoring orthodoxy in Alexandria and proposed an Alexandrian deacon named Dioscorus as patriarch, and on this there was long discussion, but at last Justin made no appointment and the Alexandrians elected Timothy III.

After Severus had left Antioch an orthodox candidate Paul was appointed patriarch and he proceeded to enforce conformity with the decrees of Chalcedon. But there were many there who refused to conform or to recognize the authority of

Paul, and these seceded from the church so that the anti-Chalcedonians now became a distinct sect, refusing communion with the Chalcedonians and declining to accept the ministrations of conforming clergy. This was a definite step away from the church.

There is some obscurity about Severus' experiences in Egypt. At first apparently he was a fugitive and assumed a disguise, living in danger of being arrested and sent back for punishment. Perhaps his life as recorded in his " Conflicts " by Athanasius of Antioch, extant in an Ethiopic version edited by Goodspeed in *Patr. Orient. IV*, with fragments of a Coptic version which has passed through an intermediate Arabic one (ed. W. E. Crum, in *Patr. Or., IV*, 578–90), rather exaggerates his sufferings and difficulties : it is the usual tendency of lives of the saints to dwell much on the sufferings they had to endure. Before long, apparently, he was welcomed and honoured by Timothy III, and was generally regarded in Egypt as a great church leader, the patriarch himself falling into the background. It was Severus who consecrated the great church of St. Claudius at Siut (Assiout) and delivered there a sermon still extant in Coptic, Constantine Bishop of Siut at the same time delivering an oration of welcome from which it is apparent that Severus was then recognized as the great leader of the faithful. (These texts in the Pierpont Morgan MSS., xlii (47).)

But the presence of the refugees in Egypt had its disadvantages. They were not all in accord and very soon it became obvious that the anti-Chalcedonians, who now began to be called Monophysites by the orthodox, were divided amongst themselves. Peter Mongus and his party had belonged to the more moderate section which was willing to accept the Henoticon, and that section was predominant in Alexandria, so Alexandria was left in peace. But Severus was of an extremer section and moreover was violent in the expression of his views. Both he and Julian of Halicarnassus were writers, and this brought their teaching before the community generally. Then it appeared that they differed materially. Severus held that the human body of Christ was subject to human defects, which was the orthodox view. But Julian pressed Monophysite doctrine to its logical conclusion and held that the union of the two natures in Christ made his body free from every human infirmity, so that he was immortal and impassible from the union which took place at the incarnation.   From this it followed that the passion caused no pain, it was merely an appearance of *phantasia*, a view which led to Julian and his followers being known as Phantasiasts.   To explain his views Julian compiled a " tome ", a book of which he sent a copy to Severus and other copies to various Egyptian monasteries which embraced his teaching cordially.   Then Severus wrote a refutation of the tome and it became clear that the Monophysites were divided into at least three discordant sects. In this dispute the patriarch Timothy took no part. He preferred to remain in the background and hoped that time would heal the differences and even reconcile the sectaries with the Catholic Church. With this end in view he attended a conference at Constantinople in 533, but terms were not arranged.   A second conference was planned for 535, but he died on 7th February of that year as he was preparing to go to the meeting.

Meanwhile Justin had died and the imperial throne had passed to Justinian (1st August, 527) whose policy followed the same lines as that of Justin, but was more moderate in application. Justinian was sincerely anxious to restore unity in the Church, but does not seem to have appreciated the problems which separated the several sects and parties. His policy was to conciliate, but Severus refused to be conciliated. The beginning of the new reign was a welcome relief to the Monophysites. Justinian, it is true, made severe laws for the punishment of heresy, but those laws were kept in reserve : he was too prudent to put them into operation. His wife, the ex-dancer Theodora, was openly pro-Monophysite. Perhaps she had her own views, or perhaps, as many supposed, her attitude was a piece of astute policy on the part of the emperor who did not want to drive the Monophysites into open revolt.

At Timothy's death, the Alexandrian synod met at once to elect a new patriarch, and the court eunuch Calotychius, acting on instructions from Constantinople, induced them to choose the deacon Theodosius, a moderate Monophysite and a friend of Severus. On the same day Theodosius was consecrated and at once proceeded to carry out the funeral of his predecessor, as was the established practice in Alexandria. But the people of Alexandria, stirred up by the extreme Julianists, would not have Theodosius and a new meeting of the synod elected the archdeacon Gaianus, who was induced to accept office with some difficulty, and he was then consecrated in the private house of one of the clergy. This was the more remarkable because he had actually assisted at the installation of Theodosius. Gaianus was soon expelled by the secular authorities, with much rioting and several murders. But Theodosius could not venture to appear openly in the city, he had to remain outside in the monastery of Canopus.

In the course of this same year (535) there was a new patriarch at Constantinople, Anthimus who, though not a Monophysite, was very much inclined towards them. By now a number of deprived Monophysite bishops, including several of the extremer section, were in Constantinople as guests in Theodora's palace, a thing which caused great scandal to the orthodox.

About this time another figure came forward. That was *Sergius of Rashayn* (c. 536) a celebrated physician and philosopher, skilled in Greek and the translator into Syriac of various works on medicine, philosophy, astronomy, and theology. In the life of the Nestorian Catholicos Maraba there is a reference to a certain Sergius who is described as an " Arian " with a tendency to paganism whom Maraba said that he would like to meet for a discussion and perhaps bring him to the true faith. No doubt this was the Sergius mentioned. In 535 he went to Antioch to lodge a complaint against a bishop named Asylus. But Ephraem, the Patriarch of Antioch, was himself in an uneasy position. He was the orthodox patriarch and had been prominent as a persecutor of the Monophysites. Now the Monophysites seemed to be in the ascendant under the protection of Theodora and he feared the possible restoration of Severus to the see of Antioch. Observing that Sergius was a man of learning and culture and familiar with Greek he sent him to Pope Agapetus to enlist his support in an appeal to the emperor to use stricter measures against the Monophysites. Sergius found Agapetus on the point of starting for Constantinople on a different errand, to obtain terms of peace for Theodahad who wished to be reconciled with Justinian. The Pope and Sergius travelled to Constantinople together. Agapetus did not succeed in checking the punitive expedition preparing to deal with Theodahad, but did remonstrate with the emperor about the way in which the Monophysites were tolerated. It was not long after this that Sergius died, though our information about his life and chronology is scanty. He is generally claimed as a Monophysite, though the translations which he made from the Greek were used by Nestorians and others as well. The Syrian historian 'Abdisho' (*B.O.*, iii, 87) claims him as a Nestorian because several of his works are dedicated to Theodore who became Nestorian Bishop of Marw in 540. But Theodore of Marw was his pupil and no doubt it was on this account that he had these works dedicated to him. Certainly the Nestorian Catholicus Maraba did not count him as one of his flock. He made his appeal to the orthodox patriarch of Antioch and acted as his envoy. But there was no one else to whom he could appeal, the Monophysite patriarch Severus being in exile. There is, no doubt, a possible solution, that he changed from one religious community to another. He was not well esteemed for his moral character and this, in view of the methods on which religious controversy was then conducted, rather suggests that he was a convert from one sect to another. Or, it may be, that he was a man indifferent to these sectarian differences and having regard only for his own career. In his earlier days he had attended the school of Alexandria and used his familiarity with Greek to prepare Syriac translations of the leading authorities studied there. As cited by Hunayn ibn Ishaq in his *Risala*, these translations covered the chief part of the Alexandrian curriculum, though that had not at that time taken its final form. Two treatises of Galen were added to that syllabus later, *De sectis* and *De pulsibus ad Tironem*. These he did not translate, their Syriac versions were made by Ibn Sahda in Muslim times. Hunayn ibn Ishaq describes them as very poor translations, but Hunayn's standards were exceptionally high. A good deal of what survives of Sergius' work is preserved in Brit. Mus. Add., 14658.

The result of Pope Agapetus' intervention was that measures were taken against the Monophysites. A synod was held at Constantinople and both Anthimus of Constantinople and Timothy of Alexandria were deposed, whilst Severus was formally anathematized. A new patriarch Mennas was appointed to Constantinople. After this experience Severus

retired again to Egypt where he died. The exact date of his
death is not known but is given variously as 538, 539, 542, or
543. He left many works, but of these only Syriac translations,
mostly fragmentary, survive. His great achievement was that
he definitely formulated the Monophysite creed. Decidedly
opposed to the decisions of Chalcedon and equally unwilling
to accept the Henoticon, he was careful not to accept the
extremer doctrine of Eutyches or that of Julian of Halicarnassus,
indeed in many respects seems to come nearer the doctrine of
the Catholic Church than would be expected of a Monophysite.
It would seem that, as the controversy first began with Eutyches,
and as Julian was the noisier controversialist, their extreme
views have often been assumed to represent the Monophysite
faith. But Severus taught a more moderate doctrine. Still he
and his followers must be classed as schismatics, if for no other
reason than that they refused to accept the considered decisions
of the Council of Chalcedon.

(4) ORGANIZATION OF THE MONOPHYSITE CHURCH

The death of Severus of Antioch marks the close of another
period of the history of the Monophysites. Now as the result of
Severus' labours they had a definite corpus of doctrine stated
in clear terms, though not as yet accepted by all sections of the
Monophysite community. But they were a community without
organization. Their bishops deprived of their sees were unable
to ordain new priests, and in many parts their adherents had
to go without the sacraments because clergy were lacking and
they refused to accept the ministrations of the " Chalcedonian "
clergy. The decrees of Chalcedon were strictly enforced by
Justin, less strictly by Justinian. But the Empress Theodora
was the mainstay of the Monophysites and several of the
deprived bishops were maintained as pensioners in her palace.
The orthodox patriarchs of Antioch, especially Euphrasius
(521–6) and Ephraem (526–46) were vigorous persecutors of
the Monophysites of Syria. A certain monk of the convent on
Mount Izla, *Ya'qub of Tella*, commonly known as *Ya'qub
Burde'ana* or " Ya'qub of the horsecloth " in allusion to the
coarse garment he usually wore, greatly distressed at the
troubles of his fellow-Monophysites, went with a monk of
Tella named Sergius to the city of Constantinople to plead their

case before Theodora. He stayed in Constantinople fifteen years protected by Theodora who showed him much sympathy, but could at the moment do nothing more. Then in 543 Harith ibn Jabala, king of the Arab tribe of the B. Ghassan which was subsidized by the Byzantine government to protect the Syrian frontier and whose chieftain was formally granted the title of " king " by the imperial government, arrived at court and asked Theodora to arrange for some bishops to be sent to the Arabs of Syria. At Theodora's request Theodosius, the exiled Patriarch of Alexandria who was living as a pensioner in her palace, consecrated a certain Theodore as Bishop of Bostra, the great mart on the Syrian frontier where merchandise from India and Arabia brought overland by the trade route from Yemen, through Mecca and the Hijaz, had to pass the imperial customs, and at the same time consecrated Ya'qub Burde'ana Bishop of Edessa. This was merely a titular dignity, as it was understood that he was to serve as a travelling bishop organizing the Monophysite community in Syria and Asia Minor, whilst Theodore did a similar office to the Arabs of the frontier and in Arabia. Of the two, Ya'qub was the more efficient : he travelled through Syria, Asia Minor, Egypt, and other parts, always in disguise and with a price on his head, everywhere organizing the Monophysite community as an independent church, consecrating bishops, ordaining priests, and supervising the administration, so that he is justly regarded as the real founder of the Monophysite Church, which is commonly called " Jacobite " after him. In 542, or perhaps in 539, his friend Sergius had been appointed (Monophysite) Patriarch of Antioch. There was an orthodox patriarch whose name appears in the official lists, but Sergius was the one recognized by the Monophysites or Jacobites. The dignity was merely titular, as no Monophysite bishop was allowed to live in Antioch. Unfortunately the Monophysite community was disturbed by many internal dissentions, which Ya'qub was not able to allay, though they caused him much vexation. In 578 he set out for Egypt to confer with Damian the Patriarch of Alexandria about these difficulties, but was taken ill on the way and died in the monastery of Mar Romanus.

Although the Monophysite Church was not organized and fully equipped as an independent body before the time of Ya'qub Burde'ana there had been several brilliant leaders already in Syria, amongst whom Ya'qub of Sarug and Philoxenos were the most prominent.

Ya'qub of Sarug who was periodeutes or rural bishop of Haura in the diocese of Sarug about 502–3, translated to the see of Batnan in the same district in 519, and died in 521, has left many letters, most of them in the manuscripts Brit. Mus. Addit. 14587 and 17162, but his fame rests chiefly on his poetical compositions, especially his metrical homilies, which had many imitators.

Philoxenos, in Syriac Aksenaya, was an alumnus of the school of Edessa where he had been trained under Hibha, but belonged to the anti-Nestorian minority which held out against Nestorian teaching. It is said that it was he who prompted Bishop Cyrus to persuade Zeno to close the school of Edessa in 489. In 485 he was consecrated Bishop of Mabboug (Hieropolis) by Peter the Fuller of Antioch. He visited Constantinople in 499 and again in 506, each time suffering a good deal from hostile officials, and in 512 presided over the synod which elected Severus to the patriarchal see of Antioch. But at Justin's accession he, with 53 other leading Monophysite bishops, was sent into exile. He went to Philippopolis in Thrace, then to Gangra in Paphlygonia and there he was murdered in 523. He was the author of a number of homilies in prose, theological treatises, letters, and several forms of liturgy, but his fame rests chiefly on a new and revised version of the Syriac New Testament prepared under his direction by his chorepiscopos Polycarp and finished in 508. Part of this version was published in England by Pococke in 1630, but an inaccurate manuscript (now in the Bodleian) was used. A phototype edition of another manuscript of this version from a codex in private possession in America was published by Isaac H. Hall in 1888, but the whole text is not accessible, though several times it has been reported as discovered. For some time this revised translation was in great repute, but the Monophysites afterwards produced improved versions which superseded it.

Mara (d. 527) Bishop of Amid was one of those expelled from his see by Justin in 519. He was sent into exile with Isidore Bishop of Kennesrin to Petra in Arabia. At Justin's death in 527 he was allowed to go to Alexandria where he spent the remaining years of his life. In Alexandria he procured a copy of the gospels and to this text he composed a prologue in Greek. All these instances illustrate the intellectual activity of the Monophysite community.

A prominent Monophysite leader was *John bar Cursus* (d. 9th February, 538), Bishop of Tella (Constantina), who was consecrated in 519, one of his consecrators being Ya'qub of Sarug. In 521 he was deposed by Justin, but went to Constantinople to plead his cause. On his way home he was arrested by Ephraem the patriarch of Antioch, a great persecutor of the Monophysites, and cast into prison in the monastery of the Comes Manasse. There he died in 538. Much of his life was spent in Monophysite propaganda along the Syrian border and amongst the neighbouring Arab tribes. He has left a collection of canons, " Quaestiones," and some other prose books.

Contemporary with him was *She'mon* Bishop of Beth Arsham, near Seleucia, who was consecrated under the Catholicus Babai (498–503), and died in 548. He was a student of the Aristotelian logic and an indefatigable controversialist who, like John bar Cursus, laboured to extend Monophysite doctrine. He travelled about Persia and Mesopotamia rallying the Monophysites and holding disputations with Nestorians, Eutychians, and Manichaeans, earning thereby the title " the Persian Disputant ", one of the few vigorous advocates of Monophysitism in Persia. Some time towards 503 he was made bishop of the small see of Beth Arsham, near Seleucia. He visited the great Nestorian stronghold of Hira several times and went three times to Constantinople to consult with the Empress Theodora. During his third visit he died. Of his letters only two are extant, one a strongly prejudiced account of the rise and spread of Nestorianism with derisive remarks about many of the Nestorian leaders ; the other on the persecution of Christians in Najran in Arabia by the Jewish Yemenite king Dhu Nuwas in 523, a persecution which is supposed to be the subject of Qur'an 84.

Another Monophysite advocate was *Isho'* (*Joshua*) *the Stylite*, originally a monk in the monastery of Zuqnin, near Amida. He wrote a chronicle of the Persian War which is our best authority for that period, but shows a Monophysite bias in the way characters are selected for admiration. This chronicle was written about 515 (ed. Martin, *Chronique de*

Josue le Stylite, 1876, in *Abhand. für d. Kunde d. Morgenlandes*, VI, and W. Wright, *The Chronicle of Joshua the Stylite*, composed in Syriac, with trs. and notes. Camb., 1882.)

The hymn writer *Shem'on Quqaya* (the Potter), of Gershir, near the monastery of Mar Bessus, composed hymns as he worked at his potter's wheel. Ya'qub of Sarug heard about him from the monks, visited him, took away some of his hymns, and encouraged him in the exercise of his poetic gifts. " A specimen of these ḳūḳāyāthā has been preserved in the shape of nine hymns on the nativity of our Lord, Brit. Mus. Add. 14520, a MS. of the eighth or ninth century " (Wright, *Hist. Syriac Literature*, 79).

One of the prelates who suffered under Justin was *John of Aphtonia*, abbot of the monastery of St. Thomas at Seleucia. He was expelled from his monastery, but founded another at Kennesrin (Qen-neshre), in the neighbourhood of Edessa. This new foundation flourished at the beginning of the seventh century for teaching Greek and was frequented by many Monophysite scholars. The Monophysites never developed an academy like the Nestorian foundations at Nisibis and Jundi-Shapur, but this monastery became quite as much a centre of scholarship.

John of Ephesus, or of Asia, was a Monophysite monk who had to flee from his monastery to escape persecution and took refuge in Constantinople in 535. There he met Ya'qub Burde'ana. He was in favour with the Emperor Justinian, who employed him in the imperial service and sent him to Asia Minor to preach amongst the pagans still to be found round Ephesus. But when Justinian died he had a troubled life. The date of his death is not known, but he was alive in 585. His official title was " Bishop of Ephesus over the heathen ". He is of interest chiefly as the author of an Ecclesiastical History in three parts : the first two parts, each in six books, cover church history down to the year 572, the third part, also in six books, carries the history down to 585, covering the period of which he had personal knowledge and as he had contact with Ya'qub Burde'ana and other leading Monophysites, this contains material of great value. Much of the work exists in a fragmentary form, but many of the fragments are of considerable length. Most of it is contained in Brit. Mus. Add. 14640, which was edited by Cureton in 1853.

G

Of this an English translation was published by Payne Smith in 1860, and a German translation by Schoenfelder in 1862.

John of Ephesus' history is supplemented by the Greek history of Zacharias Rhetor (or Scholasticus), of the later sixth century. Unfortunately this work is not extant, but there is a sixth century compilation in twelve books by an anonymous Monophysite containing material gathered from various sources, books 3 to 6 giving the greater part of Zacharias' history, covering the years 450–491. The original work seems to have gone down to 518, and the Syriac translator was writing as late as 569, or even later. This history, surviving only in part in its Syriac version, is preserved in Brit. Mus. Add. 17202.

(5) PERSIAN MONOPHYSITES

Ya'qub Burde'ana never worked in Persia, but about 559 he consecrated Ahudemmeh as bishop of Tagrit in the high-lands of Adiabene, a district which had steadily resisted Barsauma and the Nestorians and became the focus of Persian Monophysitism. Ahudemmeh proved himself a vigorous missioner who did much to spread Monophysite doctrine. He even made converts of some members of the royal family and baptized one of the sons of King Khusraw I, giving him the name of George. But for this he was cast into prison and there executed in 575.

After Ahudemmeh's execution the Monophysites had no bishop in Persia until 579 when one was appointed in the person of Qamisho' who is described as " doctor of the new church built for the edification of the orthodox near the royal palace " : these are the words of Bar Hebraeus (*Chron. Eccl.*, ii, 101) who, as a Monophysite himself, uses the term "orthodox" to denote members of his own communion. It is interesting to know that the Monophysites had built a new church close by the royal palace.

In Adiabene, where Monophysite teaching had its readiest welcome, the chief centre of Monophysite activity was the monastery of Mar Mattai, probably in the place now known as Holwan on Jebel Maqlub, about four hours' journey from Mosul, in the area between the Tigris and the Greater Zab. From the time of Ahudemmeh the Monophysite metropolitan, though titular bishop of Tagrit, resided in this monastery, secure in his mountain retreat, until about 628 when Athanasius surnamed " the Camel Driver " (Monophysite Patriarch of Antioch), summoned the Persian bishops of his communion to Syria to discuss measures to be taken to promote the spread of Monophysitism in the areas where the majority of Christians had drifted into Nestorianism. Five bishops attended, amongst them Christopher, the metropolitan of Tagrit,[6] and he, on returning from Syria, removed his residence from the monastery of Mar Mattai to the city of Tagrit itself. But the honorary title of metropolitan was preserved for a bishop resident at Mar Mattai, though it was a mere compliment, all real authority being in the hands of the Bishop of Tagrit, now resident in his titular see. In 640 Marutha, a member of the monastery of Mar Mattai, was raised to the bishopric of Tagrit, and he and his successors assumed the title of " Mafrian " which thenceforward was used to denote the supreme head of the Monophysite Church in Persia and Asia generally. By this time the Monophysites had spread well to the east, and the Patriarch Athanasius was asked to consecrate bishops for those remoter parts, but this he refused to do, preferring that the eastern Monophysites should organize themselves under the Mafrian as an independent body, so Marutha created the see of Herat in Khurasan, and other oriental sees were added later (Bar Hebraeus, *Chron. Eccl.*, ii, 121).

The great centres of Monophysite scholarship were the monasteries of Mar Mattai, Tur 'Abdin on the upper Euphrates which claimed to be the oldest monastery in Mesopotamia, and Kennesrin (Qen-neshre), near Edessa. Several metropolitans were alumni of this last, Athanasius I (d. 630–1), Athanasius II, of Balad (d. 685), and others.

The strong Monophysite element in Egypt attracted a number of Syrian Monophysite monks and scholars to Alexandria to study, amongst them Paulos of Tella and Thomas of Harqel in the early years of the seventh century. H. Evelyn White (*Monasteries of the Wadi 'n Natrun*, ii, 319 *sqq.*) shows that there was a colony of Syrian monks in Scetis already in 576, and probably their monastery there, from which many valuable Syriac manuscripts have been obtained, was founded,

[6] See note on p. 186.

or purchased from the Copts, about 710 by a certain Marutha ibn Habbib. In the sixth to seventh centuries the Patriarch of Alexandria was living in the Wadi n-Natrun.

This close contact with Egypt and especially with Alexandria promoted the spread of Alexandrian teaching amongst the Syrian and Persian Monophysites. In this connection two leading characters are of particular importance.

John Philoponus of Alexandria (*circ.* 568), was for some time a Monophysite, then turned to the doctrine known as Tritheism, which had been taught first by John Ascusnaghes and for some time was the acknowledged leader of the sect which followed that teaching. Before he became a Tritheist he had written a treatise called *Diaitetes* or Arbiter at the request of Severus of Antioch, from which a citation made by St. John of Damascus survives, but the whole work is extant in a Syriac translation, obviously well received in the Monophysite community (cf. Brit. Mus. Add. 12171). He also composed a commentary on Porphyry's *Isagoge*, and this was generally adopted amongst the Monophysites as a recognized textbook. In 568 he published a criticism on a cathetical discourse by John, Patriarch of Constantinople, but the exact date of his death is not known.

With this contact with Alexandria must be associated also the introduction into Syria of the medical *Pandects* or *Syntagma* of the Alexandrian Monophysite physician Aaron, a compilation which circulated in a Syriac translation amongst Monophysites and Nestorians and became a favourite manual of medicine. As such it exercised a good deal of influence on the medical teaching at Jundi-Shapur and finally on the earlier Arab physicians. This we conclude from the fact that the later Syriac and older Arab medical writers quote freely from it.

The Arab conquest of 632 did not check the religious or intellectual life of either the Nestorian or Monophysite community. The Arabs exacted tribute, but so had the Persian and Roman governments. The tribute-paying communities were left free to follow their own laws, religion, and customs, and to lead their own cultural life. Intercourse between Egypt, Persia, and Syria was easier than before, and this favoured intellectual culture which looked to Alexandria for guidance, though as Alexandria became immersed in commercial interests that guidance had to be sought in other cities which became its cultural heirs.

The most distinguished Syriac scholar of this later period was *Severus Sebokht* (d. 666–7), Bishop of Kennesrin. He wrote letters on theological subjects to Basil of Cyprus and Sergius, abbot of Skiggar, as well as two discourses on St. Gregory Nazianzen. On Aristotelian logic he composed a treatise on the syllogisms in the Analytics of Aristotle, a commentary on the Hermeneutics which was based on the commentary of Paul the Persian, a letter to Aitilaha of Mosul on certain terms used in the Hermeneutics (Brit. Mus. Add. 17156), and a letter to the periodeutes Yaunan on the logic of Aristotle (Camb. Univ. Lib. Add. 2812). In addition to these works on logic he also wrote on astronomical subjects (Brit. Mus. Add. 14538), and composed a treatise on the astronomical instrument known as the astrolabe, which has been edited and published by F. Nau (Paris, 1899). In all this he showed himself the product of Alexandrian science and illustrated the widening scientific interests of the period. It seems that he took steps towards introducing the Indian numerals, but this was not carried on by any immediate successor. His work represents the highest level reached by any Syriac scientist and this, it will be noted, was associated with Kennesrin.

The Monophysites were diligent and successful in missionary work, travelling the deserts under the protection of the Arab tribe of the B. Ghassan. Adiabene and Beth 'Arbaye round about Tur 'Abdin already were Monophysite territory, and so Armenia and the country about Mount Izla a little north of Nisibis. Another Monophysite centre was the town of Shissar. In that town was a physician named Gabriel who was a devoted Monophysite. He was appointed chief physician to Khusraw II and at court conformed to Nestorianism which was the officially recognized form of Christianity, but reverted to Monophysitism when he saw that there was no risk of incurring royal disfavour by doing so. He and Queen Shirin, who was his patient, did all in their power to help the Monophysites and hinder the Nestorians. It is not altogether edifying to see these rival Christian bodies engaged in intrigue at a non-Christian court. Gabriel's activities were so far successful that he was able to prevent the appointment of a new Catholicus for the Nestorians when the see of Seleucia fell vacant, and so for some time the Nestorians were without an official head.

Under Justinian the Empress Theodora sent down Monophysite missionaries to Axum in Ethiopia and so secured the Ethiopians for the Monophysite Church. Ethiopia is said to have been evangelized by St. Matthew the Apostle, but the Christian religion did not penetrate inland where were many barbarous races using different languages until the days of Constantine, when Frumentius, a Christian youth wrecked on the shores of the Red Sea, began teaching some of those people the Christian faith and was afterwards consecrated Bishop of Axum by St. Athanasius. Such is the account given by Socrates (*H.E.*, i, 19), who obtained his information from Rufinus (*H.E.*, i, 9), who died in 420, so clearly there was an Ethiopian Church well established in the early fifth century.

In the days of Justinian Axum and its king occupied an important place in Byzantine politics. The emperor, sorely pressed by foes on his European and Asiatic frontiers, was no longer able to spare a fleet to police the Red Sea, and in 522 made a compact with the king of Axum, who undertook that duty as an ally of the Byzantine government. Before long the king of Axum began trying to extend his control over the coast of South Arabia, for which he had a reasonable pretext. Control of both shores was necessary for putting down piracy, the people on both shores were akin, and formerly both had been under one ruler.

The Ethiopians successfully established themselves on the Tihama, the low lying coast country, but failed in an attempt to take Mecca. How long their occupation of the Tihama lasted is not known, but the attempt on Mecca is supposed to have been made about the time of Muhammad's birth, which may have taken place in A.D. 570 or thereabouts. The attempt on Mecca failed, but the Ethiopians were good warriors and many of the princes of South Arabia purchased Ethiopian slaves as suitable recruits for a body-guard. This example was followed at Mecca. The Meccan merchants seem to have been an unwarlike people, relying much on mercenaries for the defence of their city and on occasion armed their Ethiopian slaves as a defence force, but did not trust them very much as in time of peace those slaves were harshly treated and many ran away. A number of such fugitive slaves escaped when Muhammad was in Medina and rallied round him there, for he had already shown his sympathy for them. In his time there were many such slaves in Mecca, and many Ethiopian craftsmen, a proportion of whom probably were ex-slaves, all men of humble rank and mostly Christians and of the Monophysite communion. It was commonly said that it was from these that the Prophet learned the bible stories which figure so prominently in the Qur'an. Opponents said that " he is taught by others " (Qur. 44, 12) that " a certain one teaches him, . . . but the tongue of him whom they suggest is foreign, whilst this is pure Arabic " (Qur. 16, 105) : it was stated that this foreign mentor was one of those who came hither by violence or fraud (Qur. 25, 5), which clearly hints that he was an Ethiopian. But these humble Christians of Mecca were an unorganized community, they had no church and no bishop (cf. H. Lammens, " Les chrètiens à la Mecque à la veille de l'Hégire," in *L'Arabie occidentale avant l'Hégire* (Beyrouth, 1920, pp. 47–9). Such an origin would explain the looseness and inaccuracy of the bible stories as they appear in the Qur'an.

The city of Najran in Arabia, not far distant from Mecca, also was Christian and Monophysite (cf. H. Lammens, " La Mecque à la veille de l'Hégire," Beyrouth, 1924, pp. 256–7, 289–90). It is not possible to identify a Monophysite centre for the transmission of Greek culture to the Arabs with the same assurance as the Nestorian medium at Jundi-Shapur can be identified, but this contact must not be ignored. The Monophysite centres of learning, it is true, were monasteries, not academies like Jundi-Shapur, and so not so intimately in touch with the Arabs as the Nestorian school, but there was in contact, as appears from the fact that the mysticism of the pseudo-Dionysius and Hierotheus was brought to bear on the formation of Muslim philosophy. But a great deal of pro-Greek influence came to Baghdad through Marw and, bearing in mind how Marutha extended the Monophysite episcopate to those eastern parts, it seems probable that a Monophysite element played its part through Marw, even though there was also a Nestorian bishop there.

CHAPTER VII

INDIAN INFLUENCE—THE SEA ROUTE

(1) THE SEA ROUTE TO INDIA

GREEK influence came to the Arabs not only directly through Syria and Egypt, but also indirectly from the east by way of India and thence through Persia. In this rather more involved line of transmission three distinct phases may be noted.

(i) To passage to India of Greek scientific teaching by the sea route leading from Alexandria to north-west India and the fuller development of that knowledge by Indian students, the results transmitted to the Arabs in the early days of the 'Abbasid khalifate in the later half of the eighth century. This was especially associated with the city of Ujjain, the Indian depot of the sea route from the Red Sea. A sea route also reached south-western India, but there were no scientific results there.

(ii) The existence in Central Asia of a focus of Greek influence in Bactria, Sogdiana, and Ferghana, surviving from the days of Alexander's invasion which, though politically wrecked by the barbarian invasions shortly before the Christian era, retained a Greek tradition and was able to spread a certain measure of Hellenism into India and the Far East. This was an area in which the Persian wars planted many captives, especially about the city of Marw, and from that city came a pro-Hellenic influence which contributed materially to the introduction of Greek science into Baghdad.

(iii) The influence of Buddhism which, although declining in India in the centuries immediately preceding the coming of Islam, had certainly prepared the ground for intercourse with the western world, and was directly responsible for the prominence of the Barmakid family, the leading patrons of Hellenism.

At an early date there was intercourse between India and the great empires of what is now called the Near East. The first traces of this occurs in inscriptions of the Hittite kings of Cappadocia in the fourteenth to fifteenth centuries B.C. Those kings bore Aryan names and worshipped Aryan deities,

and apparently were akin to the Hindus of the Punjab. Blocks
of Indian teak were used in the temple of the Moon at Ur and
in Nebuchadnezzar's palace, both of the sixth century B.C.,
and apes, Indian elephants, and Bactrian camels figure on
the obelisk of Shalmanesar III (860 B.C.). These may have
been brought by land or carried by sea. The *Rig Veda* makes
allusions to voyages by sea, and many such allusions occur in
Buddhist literature, both of rather later date but bearing
testimony to an old tradition. Sea trade no doubt came from
a port near the mouth of the Indus and passed to the Persian
Gulf, coasting along Gedrosia. The Persian Gulf was cleared of
pirates by Sennacherib in 694 B.C., and it may be assumed
that the presence of pirates implies a sea trade which increased
after the pirates disappeared. In the later seventh century it
is said that the trade of the Persian Gulf was in the hands of
the Phœnicians, who had settled in the marsh lands of the
Tigris-Euphrates (Shatt el-Arab) after their earlier homes had
been destroyed by earthquake (Justin, 18, 3, 2). Strabo refers
to Phœnician temples on the Bahrein Islands near the mouth
of the Persian Gulf (Strabo, 16, 3, 3–5), and remains of such
temples have been found and explored.

The sea route connecting the western world with India
had been known to the Greeks long before the Christian era,
perhaps before the days of Skylax, the friend and neighbour
of Herodotus, certainly before the time of Nearchus and
Alexander, as Nearchus was able to get a guide from Gedrosia
who knew the coast as far as the Gulf of Ormuz (Arrian,
Indica, 27, 1), beyond which the Arabs had a monopoly.
The course was to send goods by land to Seleucia on the
Euphrates or to Zeugma, and down the river, but the route
to the Euphrates from Antioch involved a troublesome and
often dangerous crossing of the desert, thence by river to
Charax (Mohammarah) at the mouth of the Euphrates,
thence by the Persian Gulf and along the southern coast of
Gedrosia to Patala (Haiderabad in Sind) on the lower
Indus.

The Persian Gulf later was avoided because of the anarchy
in Syria when the Seleucids lost control, and the hostility of
the Parthians, through whose country Indian goods brought
to the Persian Gulf would have to be carried. This gave an
opportunity to Arab traders. Indian merchandise could be

landed at one of their ports, Aden, etc., on the coast of Yemen, or passed to the Egyptian merchants who traded in the Red Sea. In the days of Agatharchides (*circ.* 116 B.C.) Egypt obtained Indian goods from Arab merchants at Aden or Muza, but the Egyptians had only vague notions of the way those goods were brought from India to Arabia (cf. *Periplus*, 26). Agatharchides himself evidently had no direct knowledge of the route between India and Arabia : there was no *direct* trade with India. It was quite the exception that Eudoxus twice made the whole journey by sea from Egypt to India.

Merchandise landed in Yemen was carried by land through the Hijaz to Petra. The Ptolemies tried to divert this and get Indian merchandise through the Red Sea to an Egyptian port, but they made no effort to intervene in the voyage between India and Arabia. To develop the Red Sea route Ariston was sent to explore its shores, and as a result ports were made along the Red Sea coast. Ptolemy Philadelphus (285–246 B.C.) tried to bring trade to the canal of Sesostris connecting the Gulf of Suez with the Nile and founded the port of Arsinoe (Suez) at its outlet to the sea, but this had to be abandoned owing to the difficult navigation of the Heropoolite Gulf (Strabo, 16, 4, 6), which caused merchants to prefer Leuke Kome or Aelana, both communicating with Petra and not with the Nile valley. Then he founded Berenice, which communicated with Coptos on the Nile by overland route 258 miles long. In 247 he founded Myos Hormos, 180 miles north of Berenice, with safer harbour and a shorter journey to Coptos. But the Red Sea also had its difficulties as it was infested with pirates until Ptolemy Euergetes (246–221 B.C.) stationed a fleet there to put down piracy (Diod., 2, 43, 4).

When merchandise was landed at Yemen it was brought up by land through the Hijaz to Dedan (al-'Ula), the road at one time perhaps passing through Yathrib (Medina). But in the sixth to seventh century A.D. it avoided Yathrib and on it was formed the station of Mecca, possibly after the decline of Petra, which followed Trajan's incorporation of Nabataea in the Roman Empire. The Prophet Muhammad was invited to Yathrib to act as leader of the Arabs settled there and enable them either to plunder the caravans passing up from Mecca, or perhaps divert the caravan route to Yathrib.

In his days the route certainly did not pass through Yathrib. This route through the Hijaz was the famous " incense route " by which the incense of South Arabia was carried. The incense, chiefly myrrh, frankincense, cassia, and spikenard, really was the produce of Arabia, and had been purchased from the Arabs by the Egyptians, Babylonians, Jews, and others. No doubt this was a lucrative trade, but it hardly suffices to account for the exaggerated estimate of the wealth of Arabia given by Greek and Latin writers. In speaking of that wealth those writers apparently reckoned all the merchandise procured from Yemen, though in fact a great deal of this was the produce of India, some of it from Somaliland, the South Arabian ports being merely depots of transit where this produce changed hands. As the western world, at least until well into the first century A.D., received the bulk of it from Arabia, it was commonly reckoned as Arabian. Akin to this was the fact that India and Arabia were long confused, so that we cannot be sure in legends of apostolic missions whether the apostles concerned were supposed to have gone to India or to Arabia. It was a very old confusion, based on the idea that tropical Africa extended beyond the southern seas and connected with India. Thus Aeschylus (*Supplices*, 286) groups India with Ethiopia, and probably Homer (*Odyss.*, 1, 23) referring to " eastern Ethiopians " means Indians and so implies the same. Older ideas pictured a continent spreading across from Africa to India, with Arabia as a kind of half-way house on the northern shore of the lake-like water to the south of Bab el-Mandel, and it was not until the second century B.C. that exploration showed this idea to be erroneous, and several centuries more had to pass before popular opinion admitted its error.

The course between India and South Arabia, the route already used by Nearchus and by the Arabs and Indians, was known to exist, but the Greeks knew no details about it beyond the reports made by Nearchus and Skylax : probably detailed information was deliberately kept secret by the Arabs who wished to retain their monopoly of the trade, who invented travellers' tales about monsters and perils to discourage competition. After reaching South Arabia goods might be carried overland by the Arabs to Aila or Gaza, or up into Syria, thus avoiding the Red Sea passage. The Red

Sea itself presented the problem of piracy, a difficulty with which the Ptolemies were unable to deal permanently. That sea swarmed with pirates and the coasts were peopled with savages, though these were to some extent restrained towards the south by the kings of the Homerites (Himyarites) and Sabaeans. Merchant vessels had to carry a company of archers to repel Arab pirates (Pliny, *H.N.*, 6, 101), who were greatly dreaded because they used poisoned arrows (ibid., 176).

This route does not seem to have been developed by the Romans before the end of the reign of Gaius (A.D. 40–41), then the custom arose of following the Arabian coast on the outward journey only as far as Cape Syagrus (Ras Fartek), then venturing on the open sea across the Indian Ocean to Patala. After that date men who wished to go south of the Indus took a " shorter and safer " course from Cape Syagrus directly across the Indian Ocean to Sigerus, the Melizagara of the *Periplus Maris Erithrei*, which probably was either Jaigash or Rajapur. The Romans had by this time found that they could avail themselves of the monsoons, blowing west to east for six months, then six months in the contrary direction, so that a vessel could drift to India in the season, and drift back six months later. This meant that a ship crossing from the mouth of the Red Sea would reach Malabar or some part of India farther south, and the evidence of Roman coins found in India shows that many must have done so. About A.D. 50 it began to be the practice for those desiring to go across to Malabar after leaving Arabia Eudaimon (Aden) or Cane (Hisn Ghorab) " by throwing the ship's head off the wind with a constant pull on the rudder and a shift of the yard (thus sailing in an arc) go across to Malabar marts in forty days " (E. H. Warmington, *The Commerce Between the Roman Empire and India*, 1928, p. 46). The return voyage was made by tracing a southern curve between Malabar and Cane or the coast of Arabia.

The progressive stages of this sea route are described by Pliny (*Nat. Hist.*, 8, 100 *sqq.*) in a passage which has been carefully analysed by Warmington (op. cit., 45–7). From the account given by Pliny it appears that the shorter route was made available by the direction of the monsoons, the south-west monsoon enabling a ship to make rapid voyage to India

in the summer, and an equally rapid return if it left Malabar
" at the beginning of the Egyptian month Tybis, our Decem-
ber, or at latest during the first six days of the Egyptian month
Mechir, which fall within the Ides of January according to
our reckoning : thus they arrange to return home within the
year " (Pliny, *N.H.*, 8, 104, 8). In this account Pliny shows
a great advance of knowledge since the time of Strabo. The
citation of the Egyptian months emphasizes the fact that the
Indian trade with the Roman Empire was operated from
Egypt.

The *Periplus* ascribes the discovery of the use of the south-
west monsoon for the shortening of the journey to Hippalus,
a pilot or merchant, and states that all these routes which left
the coast and crossed the ocean were suggested and planned
by him. He is not mentioned by Pliny, but the name Hippalus
is given to the south-west monsoon. The *Periplus* is a careful
and accurate book of sailing directions, but in this part must
be regarded with reserve. Did the unknown author relate a
popular legend based on the name given to the wind ? In the
Itinerarium Augusti and in Ptolemy Hippalus is used as the name
of a sea. If he were a real person, it is strange that his exploits
were so little known to succeeding generations. No doubt the
" discovery " implies the judicious use of information gathered
from mariners and so giving an idea of the lie of the Indian
coast. Nearchus knew that he had to wait for the north-east
monsoon to make the voyage from India homewards several
centuries before the supposed Hippalus (cf. Arrian, *Indica*,
21, 1). Warmington points out that Hippalus only " observed
the placing of the ports, and the shape of the sea, and appears
to me only to have realized in theory the southern extension
of India and the *possibility* of using for crossing to various
points a wind which only his successors durst fully to use in
practice by successive stages " (Warmington, op. cit., 46–7).
Pliny, writing after A.D. 51, says that only after the final
development of the discovery did a regular use of this south-
west monsoon take place " every year ", and that only of late
had reliable information about the whole voyage from Egypt
to Muziris and Nelcynda been made available (Pliny, *N.H.*,
101, cf. Warmington, op. cit., 47). The use of the monsoons
to shorten the duration of the passage to and from India
was only made known to the Romans in the days of Claudius,

and so Pliny speaks of its having taken place in his own time (Pliny, *N.H.*, 8, 101, 86).

In fact, however, the voyage to India had become familiar in much earlier times, and seems to have been first explored and used by the Indian mariners. Eudoxus had sailed to India in 118–112 B.C., the route being shown him by a ship-wrecked Indian seaman found near the entrance to the Red Sea (Strabo, 2, 8, 4). Thus the discovery made in the first century A.D. was simply that the navigation of the Indian Ocean was then first made known to the Romans. The name Hippalus was given to the wind, or to the sea, its origin unknown, and the legend of the first century mariner was invented to explain the name.

Before the age of Augustus very few Greek or Roman travellers had ventured beyond the Bab al-Mandeb into the Indian Ocean, although a good deal of trade had taken place between the western world and India. " Discoveries of coins are regulated by chance, and although they indicate commerce, do not afford conclusive evidence of its extent at any given period. . . . Hardly any authenticated Ptolemaic or Seleucid coins have turned up in India, and of Roman Republican coins only a few have been found in North-West India. . . . But of emperors down to Nero very large numbers of gold coins and silver coins have been found in the Tamil states, and of these a phenomenally large number have stamps of Augustus or Tiberius " (Warmington, op. cit., 39). This at least indicates a greatly increased intercourse with India in the time of the early emperors.

To a great extent the rarity of Greek and Roman trade at an earlier period was due to the fact that the Homerites or Himyarites, the Arabs of the south coast of Arabia, who then controlled the trade, as well as the Axumites, who were Himyaritic colonists settled on the African side of the Red Sea, desired to keep the Indian trade a monopoly for themselves and were unwilling to let any strangers into their secrets. That the Axumites participated in this trade is clear from the Buddhist monument found at Axum.

Some time about 150–140 B.C. the Mongolian tribes of Yueh-chi or Sakas invaded North-West India and overran Bactria. Gradually they settled down and a confederation of Saka states was formed which became the powerful kingdom of Kushan which lasted until A.D. 226. Under the third Kushan King Kanishka (A.D. 120–153) this kingdom was at its best and trade with the western world was active, chiefly by the sea route connecting Alexandria with India, and at the Indian end of this route, some distance inland, was the great depot of Ujjain. Kanishka was a convert to Buddhism and many Buddhist monasteries were founded in his dominions during his reign. On his earliest coins the inscriptions were in Greek script and in the Greek language, the sun and moon represented in Greek form as Helios and Selene. But later in his reign, though the Greek script was continued, the Old Persian language known as Pahlawi was used and the deities represented were mixed Greek, Persian, and Hindu, a few showing the figure of Buddha. In the Kushan capital Puru-shapura (Peshawar) there was a great tower with relics of Buddha and a large Buddhist monastery, and these buildings existed until the eleventh century when they were destroyed by Mahmud of Ghazna. The fourth Kushan King Huvishka (153–185) remained faithful to Buddhism, but his successor Vasudeva (185–226) turned to Hinduism and the worship of Siva. From his reign down to A.D. 320 Indian history is almost a blank.

Under the Kushan kings there was a close and constant intercourse with the Graeco-Roman world, chiefly by the sea route connecting with Ujjain. Roman coins came to India to pay for spices and other Indian luxuries in quantities which the Emperor Tiberius deplored (Tacitus, *Ann.*, 2, 33 ; 3, 53 ; Dio Cassius, 57, 15), a complaint endorsed by the finds of Romans coins in India. The Kushan kings were the only Indian princes who themselves issued a gold coinage at that time, and in their gold coins copied the Roman model. Roman gold circulated freely throughout India.

In the third century the Kushan power declined and was restricted to the Indus valley and Afghanistan. After the time of Marcus Aurelius (A.D. 161–180) Roman trade with India decayed and the use of the sea route almost ceased. The accession of the Sasanids in Persia in 226 put a new and vigorous Persia in place of the effete and degenerate Parthia, and this new power was unfriendly to the Romans. Diocletian endeavoured to reorganize the Roman Empire to cope with new dangers which threatened its existence, but it was not until 324 that Constantine united it under firm control, and only then was interest in eastern trade revived. But times had changed, and Constantinople became the rival of Alexandria, though the route from Constantinople via the River Euphrates and the Persian Gulf was practicable only when there was peace between Persia and Rome, which was not always the case. The sea route between India and Alexandria depended upon the safety of the Red Sea which the Romans continued to police until the days of Justinian.

In India a new dynasty appeared in A.D. 320, the Gupta monarchy founded by a raja in Magadha named Chandra-gupta, with capital at Pataliputra, like the Kushan state before it this was a kingdom in the north-west. The second ruler of this dynasty, Samudragupta (330–380) became paramount over all North-West India. He had no sympathy with Buddhism, but took a strictly nationalist attitude and adhered to Brahminism. Efforts were made to revive the Sanskrit language, and Buddhist forms in architecture became obsolete, whilst there was a great development in the form and decoration of the Hindu temples. In art, however, the Greek influence which came through Gandhara on the north-west frontier, still lingered, and the coinage, at least, continued to follow Roman models. The third king of this dynasty, Chandragupta II (380–415), extended his conquests over all Western India, subduing the country of the Sakas (Surashtru, now Kathiawar) and the Saka princes known as " the Great Satraps ". This put him in possession of Malwa and its capital Ujjain, the inland depot of the sea-borne trade with the Red Sea, and the adjacent ports Baroch (Broach), Sopara Cambay, and others. In spite of the revival of the Hindu religion, the population of the north-west remained pre-dominantly Buddhist, free from caste restrictions and without any tabu on travel.

(2) ALEXANDRIAN SCIENCE IN INDIA

Under the Gupta kings the city of Pataliputra became the home of scientific studies, especially of astronomy and mathematics, both of which show a definitely Greek impress in accordance with contemporary work in the school of Alexandria. The astronomer Aryabhata (born 476–499) taught here and has left a treatise on astronomy with a section dealing with mathematics. Varahamihisa (505–587) compiled a work known as the *Pance-Siddhanlika*, a compilation of five standard manuals of astronomy which he abridged. One of these five treatises belongs to the pre-scientific age and is of no scientific value, but the other four show the influence of Alexandrian scholarship : two of them bear the non-Indian names of *Romank* and *Paulisa*, the latter giving a table based on Claudius Ptolemy's table of chords. These treatises refer to the Yavanas or Greeks as the great authorities on science. One of the four treatises is the fifth century anonymous Surya Siddhanta or " knowledge by the sun ", which became a standard manual for Indian astronomers. Brahmagupta (*circ.* 628) was an astronomer who lived and worked in Ujjain, where there was an observatory. He wrote an astronomical manual called the *Brahma Siddhanta* in twenty-one chapters, including special sections on arithmetic (*Ganitad'haya*) and indeterminate equations (*Kutakhadyaka*). This work became known to the Arabs during, or a little before, the reign of Harun ar-Rashid and formed the basis of the work which circulated as the *Sindhind*, a name which represents the Indian *Siddhanta*.

Under the Sasanid kings of Persia it had been the custom to take and record astronomical observations, no doubt in the first place for astrological purposes, and these records were regularly published as the *Zik-i-shatroayar* or " royal tables ". The preparation of those tables was not stopped by the Arab conquest, nor were they greatly changed in form, the Persian language was still used and not replaced by Arabic for several centuries, and even then the dates were given with the old Persian months, not the months of the Arabic Muslim year. It is known that there was an observatory at Jundi-Shapur, and no doubt observations were taken there as well as in the Persian observatories, but the whole work was and remained in Persian hands. Then, apparently, the Arabs wanted to understand how these observations were taken and recorded and for that purpose the *Sindhind* was composed and circulated amongst them. It was the first astronomical manual introduced to the Arabs, and it included not only astronomical information, but also the mathematical material necessary for its use, mostly dealing with spherical trigonometry.

H

There is a legend, but it is a dubious one, which puts back the translation of the *Sindhind* to the reign of al-Mansur, the founder of Baghdad. This legend relates that the Arabs conquered Sind (Scind), the area of the lower Indus, in the days of their expansion after the fall of the Persian monarchy, which has a good historical basis. This conquest did not result in a complete occupation of the country, but certain Arab chieftains were settled there as a kind of military garrison to hold it, and they, very naturally, became semi-independent. When the 'Abbasid revolution took place they seized the opportunity to declare themselves independent and refused to recognize the new dynasty. But al-Mansur would not tolerate this and sent an armed force to chastise them, and after that experience they determined to make their submission and sent an embassy to Baghdad to make terms. With this embassy went an Indian sage named Kankah, who disclosed to the Arabs the wisdom of the Indians, which consisted of a summary of astronomy and the mathematics involved. But Kankah knew no Arabic or Persian, and his speech had to be translated into Persian by an interpreter, and that into Arabic by a second interpreter, a process which rendered the final form of his instruction very involved and obscure. Al-Biruni (d. 1048), the earliest and best Muslim observer of India and Indian things, knew this story but did not believe it and considered it an invention designed to explain why the translation of the Arabic Sindhind was so obscure and unsatisfactory. History knows of no embassy sent from Sind to al-Mansur. The probability is that the work was an Arabic translation of a Persian version of the Siddhanta already in use in Jundi-Shapur. In any case its contents are not a collection of notes of the discourse of any sage, but a translation, or rather paraphrase, of the standard Indian manual, the revised Siddhanta of Brahmagupta. There may be this much truth in the story, that the Siddhanta passed through two translations on its way to the Arabs, or possibly three, from Indian to Persian, possibly thence into Syriac, finally into Arabic.

The mathematics and astronomy which the Arabs learned from their Indian teachers through a Persian medium were of Greek origin, passed from Alexandria to North-West India. But it does not seem that the actual Greek authorities circulated in India, their teaching was assimilated and restated by Indian scientists, who developed and made material contributions to the material which passed through their hands, and rendered it more flexible by the use of a decimal notation and a greatly increased use of symbols. This can be estimated by noting the work of Aryabhata. It appears from al-Biruni that there were two scientists bearing this name (al-Biruni, *India*, ii, 305, 327). The elder of these seems to have died about A.D. 500, the date of the younger one is unknown, nor can we always distinguish which of the two is meant. The elder Aryabhata worked at Pataliputra, not at Ujjain. He produced several works, the Gitika, which was a collection of astronomical tables, the Aryashtasata, which includes a treatise on arithmetic known as the Ganita, and a treatise on the geometry of the sphere the necessary basis of astronomical work known as the Gola. He solved quadratic equations, already anticipated by Diophantus who, however, recognized only one root, even where both are positive, and had been already suggested by Heron. He attempted indeterminate linear equations, already anticipated by Hypsicles, and gives one of the earliest attempts at the general solution of such equations by means of continued fractions. He sums up an arithmetical series after the pth term in a way which may be expressed—

$$S = n\left\{a + \left(\frac{n-1}{2}+p\right)d\right\}$$

He gives rules for determining the area of plane figures, but often expresses himself very imperfectly, as "the area produced by a trilateral is the product of the perpendicular which bisects the base and half the base". He gives the area of the sphere as $\pi r^2 \sqrt{\pi r^2}$, which makes $\pi = \frac{16}{9}$ perhaps an error for Ahmes' $\left(\frac{16}{9}\right)^2$. For the value of π he says, "add four to one hundred, multiply by eight, add sixty-two thousand: the result is the approximate value of the circumference when the diameter is twenty thousand." This makes $\pi = \frac{62,832}{20,000}$ or $3 \cdot 1418$.

In his astronomical tables he includes a brief table of sines and rules for finding them. In all this there are traces of

Greek teaching, and that appears also in his terminology, as *jamitra* = διάμετρος, *kendra* = κέντρον, and *drama* = δραχμή. His work goes farther than that of the Greeks because, like other Indian scientists, he makes a freer use of algebraic expressions, which were rather tentatively introduced by diophantus, and employs the far more convenient Hindu numerals.

Brahmagupta (*circ.* 628) worked in the Ujjain observatory. He was the author of the Brahma-Siddhanta " Brahma's revised Siddhanta ", which was the basis of the Arabic Sind-hind. This work contains chapters on arithmetic and a treatment of indeterminate equations. In the arithmetic he deals with integers, fractions, progression, barter, rule of three, simple interest, mensuration of plane figures, volumes, and " shadow reckoning " or use of the sun dial. His rules for areas are often defective : thus for an equilateral triangle with side 12 he gives $5 \times 13 = 65$; for a triangle with sides 13, 14, 15 he gives $7 \times \frac{1}{2} \times (13 + 15)$ which is 96. His formula for the area of a quadrilateral with sides a, b, c, d, is $\sqrt{(s - a)(s - b)(s - c)(s - d)}$, where $s = \frac{1}{2}(a + b + c + d)$, but this is true only for cyclic quadrilaterals. His rule is expressed thus, " Half the sum of the sides set down four times and severally diminished by the sides, being multiplied together, the square root of the product is the exact area." He takes π as 3 for practical purposes, or $\sqrt{10}$ as its exact value. He deals with quadratic equations of the type $x^2 + px - q = 0$, taking $x = \dfrac{\sqrt{p^2 + 49} - p}{2}$ which gives one root correctly. More important is his application of algebra to astronomy in the Kutakhdyaka, the first instance of such an application being made. He considers simultaneous equations of the first degree, calling their unknowns " colours ". Considering the solution of $ax - by = c$, he gives $x = \pm cq - bt$, and $y = \pm cp - at$. This had been already considered by Aryabhata, who, however, did not solv: it, now Brahmagupta gives a solution. These formulae assume that t = zero or any integer and that $\dfrac{p}{9}$ is the penultimate convergent of $\dfrac{9.}{6}$ For the right-angled triangle he gives two sets of values, $2mn$, $m^2 - n^2$, $m^2 - n^2$, and $\sqrt{m}$, $\frac{1}{2}\left(\dfrac{m}{11} - n\right)$, $\frac{1}{2}\left(\dfrac{m}{11} + n\right)$, in which he probably draws from Greek sources.

For such treatment it is obvious that Indian mathematics of the period when there was a regular sea route in use between Alexandria and Ujjain were based on Alexandrian Greek teaching.

As Arab astronomy began with a continuation of the work in progress in the Persian observatories, which work was rendered possible only by the use of Indian mathematics, it seems fairly certain that the Arabs must have used this Greek science which came through an Indian medium, and was transmitted from the Indian scientists by Persian astronomers and mathematicians, although the Persian books which supplied the Arabs with this knowledge are no longer available. It is said that when the Arabs found themselves unable to understand the Almajest Ja'far ibn Yahya the Barmakid at once knew the required remedy to be a knowledge of the text of Euclid and Claudius Ptolemy, material at that time not yet accessible in Arabic. If this statement can be treated as reliable it suggests that he, a Persian of Persian education, was familiar with the needed material, though a Persian version, or for that matter an Indian one, of those two authorities is unknown. It is not necessary to prove that translations of the Greek scientists were actually made in Hindu or Persian, it is sufficiently clear that their teaching was known and used.

CHAPTER VIII

INDIAN INFLUENCE II—THE LAND ROUTE

(1) BACTRIA

INDIA could be reached by land as well as by sea. It is known that there was trade with India in Assyrian times, but it is not clear whether this was by land or sea. Direct evidence of intercourse between India and Western Asia begins in the Persian period after Cyrus broke through the hostile tribes which had hitherto barred the way. Darius, the son of Hystaspes (521–485 B.C.), penetrated into North-West India and annexed the Indus delta which thereafter was claimed as a Persian starapy, as appears from the inscriptions of Persepolis and Naksh-i-Rustam. It was this Darius who in 512–510 sent the Greek pilot Skylax, of Karyanda, in Karia, the neighbour and probably the friend of Herodotus, to explore the practicability of a short sea route between the Persian Gulf and the mouth of the Indus, which seems to imply familiarity with the Indus country. As soon as he knew that there was such a route available he sent a fleet into the Indian Ocean.

Alexander's invasion of India, which was chiefly intended to secure the easternmost province of Persia after the Persians had been conquered, took place in 327–325 B.C. Before crossing the mountain frontier of India he formed a military base which afterwards became the city of Alasanda or Alexandria Under the Caucasus, its site probably some 30 miles north of Kabul, one of the many Alexandrias which he founded. The term " Caucasus " was applied by the Greeks to what is now known as the Hindu Kush. Alexander died in 323, and at his death his kingdom, for which he left no heir, was fought over by his generals and in 312 was divided between them. In this division the Asiatic province fell to Seleucus Nicator, who founded the city of Antioch in Syria and made it his capital, relegating the extensive provinces east of Syria to the Indus to a subordinate position. He was more concerned with the rivalries between the Greek rulers along the Mediterranean coast than the affairs of the Asiatic hinterland, and left Babylon and all that had been the kingdom of Persia to deputies. Seleucus was succeeded by his son Antiochus Soter (280–262 B.C.), and he by his son Antiochus Theos (261–246), all three involved in wars with the Ptolemies of Egypt and so leaving Persia very much to its own devices. Taking advantage of this the Parthian tribes of East Persia (Khurasan) drifted away from Seleucid rule and formed an independent kingdom of Parthia about 250 B.C. This new Parthian state included a large part of the old kingdom of Persia, but by no means all that had been ruled by the ancient Achaemenid kings. About 210 B.C. the Seleucid king Antiochus III " the Great " formally recognized the third Parthian king Artabanes as an independent monarch.

These Parthian kings were not of the Persian royal family of the Achaemenids, but Scythians from Maeotis, though later a legend was circulated to the effect that their founder Arsaces had been born in Bactria. As derived from the semi-barbarous tribes of East Persia the Parthians were despised by the Persians proper and regarded as inferior species : they were the only tribe of their locality not mentioned in the sacred books of the Persians, and seem to have preserved some of the nomadic habits of the tribe from which they were descended. They made their winter capital at Babylon or Ctesiphon, this latter a camp city on the Tigris, avoiding the nearby Greek colony of Seleucia which was left more or less independent under its own Greek constitution and using the Greek language and religion. The summer capital was Ecbatana (Hamadan) or Rhagus. There was also a palace at Hecatompylos in the middle of Parthia, a city which had been enlarged and partly rebuilt by Seleucus. The sixth Arsacid Mithridates I (d. 138–130 B.C.) greatly enlarged the Parthian kingdom, and after extending its boundaries from the Tigris to the Indus assumed the title of " King of Kings ", which had been used by the Achaemenid monarchs, and was represented on his coins as carrying a bow like those old kings, and adopted the pearl-studded tiara which they had worn. The Achaemenids had been regarded as of semi-divine descent and as possessing a divine spirit emanating from the god Ahura Mazda, and so called themselves " sons of god ", and this title was now assumed by the Parthian kings as *Zag Alohin* in the inscriptions on their domestic coins, or θεοπάτηρ on their Greek coins. The Parthian kings were incorporated in the ranks of the

" Great Ones " (Μεγιστᾶνες) or higher nobles of the kingdom
and in the fraternity of the Magi or Persian priesthood, all
as had been under the ancient Achaemenids, and they and
the higher Parthian officials tried to assimilate themselves as
much as possible to the Persians, copying their dress and
manners and often adopting Persian names.

Alexander had left a number of colonies scattered over what
had been his empire, and these lasted and became sources
of Greek cultural influence.   But quite apart from these
colonies Alexander had left a prestige and cultural influence
whose effect endured for many centuries, so that the Asiatics
of the Near East looked with respect on all that was Greek.
Greek was not the official language in Parthia as it was in
Egypt, but Greek was very commonly used on Parthian coins,
though under the later kings it was so debased as hardly to
be intelligible.   The oldest coin, which is one of Vologasus I
in the time of the Roman Emperor Claudius, gives the full
title of the king in Greek, contenting itself with the king's
name abbreviated to VOL in the native Old Persian or
Pahlawi.   From about 188 B.C. onwards the royal title includes
the term φιλέλλην.   To some extent the Parthian state had
a Hellenizing character, though this Hellenism became more
and more orientalized.   National feeling was not developed
in its full form, as the ruling dynasty was generally regarded
as racially inferior, tolerated in office only because it had
been successful in liberating the country from an alien yoke,
and supported because it had proved its capacity to secure
peace and independence effectually : when it experienced
defeat at the hands of a foreign power it lost its hold and
people looked for a legitimist king of the original stock
descended from the demi-gods.

After the revolt of Arsaces, which led to the foundation of
Parthia, the lands of Bactria, Sogdiana, and Fergana drifted
out of the control of the Seleucids and a Greek kingdom was
formed in Bactria on the Indian border, though maintaining
intercourse with the Greek world.   This state lasted until
about 128 B.C., its population apparently often recruited by
fresh Greek colonists.   The city of Antioch Margiana or Marw
in Sogdiana was at the end of an important and well travelled
route from Syria and Northern Mesopotamia, and connected
with Bactra, the capital of Bactria, and with Alasanda or

Alexandria " under the Caucasus "on the threshold of India. Through all its history it remained definitely Greek, and was a centre of Greek influence until it fell before barbarian invaders.   As independent Bactria was in revolt against the Seleucid monarchs of Syria, and their rivals, the Ptolemies of Egypt, maintained an' agent at the Bactrian court. These central Asian states were intimately involved in the intrigues of the Eastern Mediterranean.

Bactria did not so much revolt as drift away from Seleucid control because the Seleucids neglected it.   About 248 Theodotus, the satrap of Bactria, made himself independent : Justin (41, 4) says that he ordered himself to be called king, but evidence of this does not appear on his coinage.   Certainly his son Diodotus or Theodotus II did so, and made alliance with Parthia against his suzerain at Antioch, a reversal of the policy of his father, which was unpopular.   He was slain by Euthydemus, the husband of the daughter of the widowed queen of Theodotus I, and when the Seleucid Antiochus III blamed him for slaying Diodotus he defended himself by saying that he was no rebel but had killed the son of a rebel (Polybius, 11, 34, 2), which shows that contemporary opinion held that Theodotus had revolted against his overlord.   In 208 Antiochus III " the Great " tried to recover Bactria for the Seleucid kingdom, but after two years' fruitless siege of Bactra Euthydemus threatened to call in the Sakas (Scythians) and pointed out the disaster which would follow the advent of these barbarians.   Antiochus desisted from his attempt and formally recognized the king of Bactria's independence.   In 190 Antiochus himself suffered a severe defeat at the hands of the Roman Scipio Asiaticus and for some time the threat of Seleucid conquest was averted.   In the following year Euthydemus himself died.

The next Bactrian monarch Demetrius had ambitions of extending his kingdom in the Indian direction, invaded India by the Hindu Kush, and in 175 occupied Pataliputra. This was but the first stage of his advance.   He then planned a great invasion of the Punjab, dividing his forces into three armies, all of which were to operate in concert.   He himself in command of the first army occupied Gandhara and Taxila. This Gandhara was known as " the second Hellas " because so thoroughly Greek and the Greek art which flourished there was destined to spread eastwards and influence the Far East. At the same time it was a " holy land " of the Buddhists, a sanctity acquired by the presence of three out of the four great Buddhist stupas there.   Buddha had never visited the country, it had no associations with his life or ministry, its holy character depended entirely on the presence of these monuments which enshrined important relics of Buddha or of his garments. The second army was entrusted to Menander, and this forthwith seized Pataliputra the capital of Sāgala (Sialkot), the chief town of the Madras, who also were Buddhists.   The third army was led by Demetrius' brother Apollodotus, who proceeded to Barygaza, which may mean Ujjain.   By these operations Demetrius held all North-West India. But the Seleucids did not abandon their hope of recovering Bactria, and in 168 Antiochus IV sent an expedition led by his general, Eucratides, against Demetrius. At the approach of the Seleucid army Demetrius ordered Menander to abandon Pataliputra and himself joined issue with Eucratides on the west of the Hindu Kush, and in this encounter the Bactrians were defeated and Demetrius slain.   Eucratides forthwith took Gandhara and prepared for the invasion of India, but waited for Antiochus, who planned himself to be the leader of the expedition which he hoped would be as glorious as that of his great predecessor Alexander. Before the invasion took place, however, Antiochus died at Gabae in 163 (Polybius, 31, 9, 11).   This unexpected event left Eucratides to rule conquered Bactria, but that was only for a brief period ; the Parthian King Mithridates intervened and secured Western Bactria for himself, and not long afterwards (in 159–8) Eucratides died. But the third invader Menander was still left and he probably ruled Sagala until 145. Most of his subjects were Buddhists who favoured the Greeks, whom they regarded as friends and saviours from the Hindus who persecuted Buddhism. Menander is described as being very well inclined towards the Buddhists, but there is no proof that he actually embraced their religion. In the Melindapanha there is a legend that he did so, and there is a Buddhist dialogue in which one of the interlocutors is " Melinda ", supposed to represent Menander. By this time, however, Buddhism was no longer expanding in Central Asia, its future lay rather in the Far East.

Greek Bactria came to an end between 141 and 128, an end brought about by the migration of the Saka (Scythian) tribes of the Yueh-chi who came from Northern China. They were, of course, Mongolian tribes, for that is the implication of the term Saka or Scythian. In China their pastures had been taken from them by another Mongol tribe, the Hiung-nu, and so they migrated, some going south where they founded a kingdom in China, others to the west where they fell upon the tribe of Wu-sun, killed their king, and occupied their lands. But before long they were overtaken by their old enemies the Hiung-nu, called in by the defeated Wu-sun and were forced to continue their march westwards. They next attacked the Sai-wong tribes who fled south, but about 160 B.C. they were themselves attacked by the Wu-sun, led by the son of their murdered king, and went farther west. Then for a while they pass out of sight until about 128 when they crossed the Jaxartes, then the Oxus, and occupied the provinces of Bactria and Sogdiana, where they founded a group of Saka states. Meanwhile the dispossessed Sai-wong had seized the Greek province of Ferghana and started another Saka principality there. The coming of these semi-barbarous tribes completely submerged the political and social life of the Central Asian Greek kingdoms, at least for the time being. It did not interfere with the Buddhist religion, for most of the invading tribes turned Buddhist.

The Yueh-chi had come from China, and the Chinese government had followed their subsequent vicissitudes and in 128 the Chinese General Chang-k'ien overtook them in Bactria and made an alliance between them and China, and for some time afterwards the Chinese endeavoured to exercise some measure of control over them, but about 48–35 they ceased to take any interest in them.

Gradually the nomad tribes settled down and shortly after 25 B.C. Kujala, chief of the Kushan tribe, one of the group composing the Yueh-chi horde, formed a Saka state in Bactria and North-West India, a combination of five older states, and this lasted for two centuries. By that time Bactria or Balkh had become a holy land of Buddhism and this sanctity was developed under the Kushan kings until Buddhist pilgrims came from many parts to visit the numerous topes or relic shrines which abounded there.

For some time Kushan Bactria is of interest chiefly as a factor in the evolution of organized Buddhism. Then it became a rising power in North-West India under King Kadphises I. Already King-hien and other Chinese scholars had visited Bactria when in A.D. 64 copies of Buddhist books were sent to the Chinese Emperor Ming-ti, with the result that in the following year Buddhism was added to the religions officially recognized in China. Under Kadphises II (A.D. 85–123) commercial intercourse with the Roman Empire, chiefly by sea rather than the land route through Marw, was greatly developed, as is noted elsewhere (above).

The third Kushan king, Kanishka (A.D. 123–153), was a convert to Buddhism. Conditions had so far changed that Kushan had checked Chinese expansion and many Chinese hostages, including Han, the son of the Chinese Emperor, were taken to Balkh. For them Kanishka built a monastery in Kapisa, and in the cold season they were transferred to a place called Chinapati, whose site is unknown. Under this king the coinage still followed a Greek model and shows a degenerate form of Greek inscription. At the Kushan court there were sculptors, trained chiefly in the school of the frontier province of Gandhara, who followed Hellenistic models. By this time Buddha was deified and worshipped, and statues representing him began to appear and take their place in Buddhist temples in place of the older allusive symbols. The earliest images were produced in Gandhara and so were designed on Greek lines, reproductions of Greek images of Apollo. Gandhara art shows Greek inspiration and carried Greek influence through the great part of the Buddhist community, so that even in China and Japan figures of Buddha show a Greek character, especially in the drapery. True to Greek standards this type of Buddha was simply a handsome man. But there were some Buddhists who were dissatisfied with this Greek type of their deity and wanted a more mystical and spiritualized figure, not a purely human form, however perfect, and so in Mathura on the great high road between Alexandria " under the Caucasus " and Pataliputra another type was devised, at first a clumsy modification of the Gandhara figure, but finally developed as a saintly and spiritualized character which, however, still betrayed its Greek origin.

(2) THE ROAD THROUGH MARW

Our main interest here is with the overland route between the Roman Empire and the Far East. That route led from the Syrian border to Marw, a city founded by Antiochus I (280–240 B.C.) as a Greek colony with surrounding agricultural settlements, all predominantly Greek, both city and rural area frequently recruited by fresh Greek colonists. Under the Parthian kings this became a mart where Roman and Chinese trade met. At the time of the Arab conquest and for long afterwards this was a scene of great prosperity, producing silk and fine cotton when those materials were still rare and costly in the Roman Empire. Before that conquest the western quarter or rabaḍ had much increased in population, and in early Arab times the main business part of the city had removed to this quarter. To Marw the last Persian King Yazdegird III fled on his defeat and there he was overtaken by the Arabs in 651 and killed at a mill in the village of Raziq close by. The Christian (Nestorian) bishop took the deceased monarch's body to Pa-i-Baban and buried it there (Tabari, *Ann.*, i, 2881), an incident suggesting that the Nestorians formed an important element in the city. There was a great Nestorian monastery at Masergasan north of the quarter known later as Sultan-Qal'a, adjoining Rabad (Tabari, *Ann.*, ii, 1925). Marw seems to have been an outpost of Hellenism, with a considerable proportion of Christians, both Nestorians and Monophysites, in its population, no doubt largely swelled by the many captives taken by Khusraw II from the Romans and sent far east for safe custody.

Marw, Bactria, and Sogdiana were all centres of Hellenism. The Saka conquest of Bactria checked, but did not destroy this Hellenic element. Meanwhile the western end of the route also had its vicissitudes. There the chief barrier between the Greek and oriental world was Parthia which was encroaching upon the Seleucid dominions and about 150 B.C. absorbed Mesopotamia. But Parthian advance was checked. Not long after the invasion of Mesopotamia came the Saka penetration of the eastern provinces. On the other hand the Seleucid monarchy ceased to be a serious obstacle when in 129 B.C. Antiochus Sidetes was defeated and slain by the Parthians, though they were not able to follow up this victory effectively because the Sakas were already menacing their eastern frontier.

This defeat left Syria too weak to defend herself from foes gathering round and only waiting for an opportunity to seize her territory. Already Arab tribes were encroaching on the eastern parts of Syria and a native dynasty at Edessa had declared its independence in 132, whilst the whole country was subject to incursions of Arab tribes who before long began preying on Parthia as well. Thus Mesopotamia became a neutral territory covered by minor native states over which neither the Seleucid king at Antioch nor the King of Parthia could exercise control.

A more formidable foe appeared in 79 B.C. in Tigranes King of Armenia, a land of hardy highlanders which had resisted Greek penetration. Tigranes easily conquered Syria, but at that time the Romans were expanding round the Mediterranean, and before long Pompey defeated the Armenians, took Syria out of their hands and made it a Roman province, with the exception of Commagene in the north-east, which was left as a vassal state under native princes. Pompey so far stabilized existing conditions as to recognize the Euphrates as a natural boundary between Parthia and the Roman Empire, though this did not prevent the Romans accepting Osrohene, with its capital Edessa, as a client state, although it was on the Parthian side of the river.

There was a chain of Arab states extending from the Armenian border to North Arabia, the most important of which was Palmyra. Augustus, who respected Pompey's recognition of the Euphrates as the frontier between Persia and the Roman Empire, seems to have regarded these Arab settlements as a kind of " buffer states " tending to protect the eastern frontier of the Empire from Parthia.

From the time of Trajan onwards the history of Western Asia centred in the prolonged duel between Rome and Parthia, or Persia, which was only Parthia reorganized under a new dynasty, and this duel had successes varying from time to time between the two combatants. The hinterland of Syria was never thoroughly Hellenized, the church councils there were conducted in Greek, but the bishops from Mesopotamia had to use the services of interpreters (Schwartz, *Acta Concil. Oecum.*, II, i, 184, 193), and the clergy of Edessa sent a petition to the Council of Chalcedon in which more than a third of the signatures were in Syriac (ibid., 35).

The Sasanid revolution of A.D. 226 placed a new Persian dynasty on the throne which had been that of Parthia. This revolution, like most such movements in oriental lands, had a religious bearing. It not only set on the throne a legitimist claimant who was accepted as descended from the demi-gods of ancient times, but it led to a drastic reformation of the religion founded by Zoroaster. The first Sasanid monarch Ardashir began his reign with a general council of Mazdean clergy which resolved the many sectarian difficulties between the various sections into which the Persian community was divided, and standardized the worship and scriptural canon. In history Mazdeanism appears generally as a tolerant creed, save in dealing with dissenters from itself, such as Mani and Mazdek, but it seems to have passed through a period of active propaganda, of which there are no details, in the course of which the religion of Zoroaster spread over the eastern provinces of the kingdom, so that at the coming of Islam Bactria, Sogdiana, and Ferghana were largely, but by no means entirely, Mazdean, with a strong Buddhist minority which proved rather a problem to the Muslim conquerors. Thus the Barmaks, heirs of the hereditary Buddhist abbots of Nawa Bahar, possessors of great wealth chiefly derived from the offerings of generations of Buddhist pilgrims, are represented as being fire-worshippers until their conversion to Islam.

The Barmakids were especially associated with the city of Marw, whither they had removed from Bactria, and they were prime movers in the 'Abbasid revolution. That revolution led to the dominance of Persian influence and to at least a partial Persianization of the Arab state, the Muslim religion, and Arabic literature. It was a Jew from Marw, Mashallah ibn Athari (d. 815–820), who was one of the astrologers called in at the foundation of Baghdad and the author of works on astronomy and mathematics which show Greek influence. It was another Jew of Marw Sahl ibn Rabban at-Tabari (c. 800) who came to Baghdad and made the first Arabic translation of Euclid's *Elements*.

CHAPTER IX

BUDDHISM AS A POSSIBLE MEDIUM

(1) RISE OF BUDDHISM

THE Hindu religion based on the cults of the Aryan invaders of India but incorporating elements from the primitive religions surviving amongst the conquered aborigines, was fully developed long before Alexander's invasion, and had evolved a rigorous caste system which divided its adherents into sharply defined and exclusive groups, raising barriers against intercourse with the outside world. But about the fifth to sixth centuries B.C. there were several religious movements, especially in North-West India, which tended to break away from Hindu ritualism, all showing a certain mystic tendency with an ascetic element and a great regard for the sanctity of human and animal life. One such movement produced the Jain religion which never spread beyond the borders of India, another was the religion of Buddha, in its earlier period a minor ascetic sect, but afterwards growing and spreading until it became one of the great world religions. Both these religions had their roots in the already existing Sankhya system of philosophy commenced by Kapila.

The Jain religion was founded by Mahavira, who preached in the kingdom of Magadha (South Bihar) in North-West India probably about 507 B.C. Gautama Buddha gathered a monastic order around him in the deer park at Sarnath, near Benares, and died about 480 B.C., but his teaching spread in the South-East Gangetic area, Kosala (Oudh), and Magadha. Thus both these religions were connected with Magadha. The whole country of Magadha was regarded as unfit for the sacrificial fire, so that no Hindu sacrifice could be offered there, and it was not a place in which Brahmans of noble and pure descent could live. This absence of Brahmans encouraged greater freedom of thought and favoured the rise of new religious views, in some measure critical of accepted doctrine (Nalinaksha Dutt, *Early Monastic Buddhism*, i, Calcutta, 1941, 140). Neither of these two religions tried to overturn the existing Hindu caste system, indeed Jains continued to employ Brahmans as domestic chaplains, but in both the laity obtained a more prominent place and caste divisions gradually lost a great deal of their significance.

In the fourth century Magadha was, it is said, ruled by kings of the Nanda dynasty ; though that dynasty of seven monarchs is often regarded as legendary, Indian political history beginning only at the appearance of the Maurya dynasty about 323 B.C., three or four years after Alexander's invasion, but it is perhaps rash to ignore entirely the legends of earlier kings. The last Nanda king is said to have been of low caste and heretical in religion, an enemy to the two higher castes of Brahmans or priests and Kshatriyas or warriors, but himself rich and powerful. There is no proof that he was a Jain or a Buddhist.

About 323–2, in the disorder resulting from Alexander's invasion, Chandragupta, of the Maurya dynasty, revolted and deposed the Nanda kings and founded an independent state. He was a man of military ability and defeated Seleucus Nicator in 305–4 ,who attempted to enforce his authority over the eastern provinces of Persia after recovering Babylon in 312. After his defeat he made a treaty with Chandragupta, recognizing him as King of Magadha (in 303), and in 301 placed a Greek agent Megasthenes at the Magadha court. Megasthenes wrote a book descriptive of India and Indian customs, which is known to us only in citations made by Clement of Alexandria and Strabo.

The next king of Magadha was Bindusara (297–272 B.C.), at whose court Megasthenes was replaced by Daimachos, who corresponded with Antiochus Soter. Both these two Maurya kings were regarded by the Hindus as upstarts and unclean, not being of priestly or warrior caste.

The third king of this dynasty, Asoka, was converted to Buddhism, which attached no importance to caste, and gave an enthusiastic support to his adopted religion. He summoned a third general Buddhist council to be held in the Asokarama in Pataliputra, a village which had been visited by Buddha at one time, and at this council eighteen sectarian differences were debated and settled and, what was of greater moment, it was decreed that Buddhism should embark on missionary enterprise and carry forward the " Law of Piety " to all the nations of the world. In accordance with this missionaries were sent out to the south and west, but not to the east. No reference to this council occurs in the Sanskrit authorities, whilst the third council mentioned in the Sanskrit books is described as having been held in Kashmir under Kanishka, this council being ignored in the Pali records which describe the council of Asoka. By these missionary efforts the island of Ceylon was converted to Buddhism of the primitive type, such as is known as Hinyana, and there are surviving records of that mission and its work. The Ceylon chronicles also refer to missionary work in the west. They state that a person named Maharakshitra led a body of missionaries to Yavana, the land of the Ionians or Greeks, but give no details of their work. At that time the Seleucid Empire extended to the Hindu Kush and politically of course all up to that boundary was reckoned as Greek. It was not until the later years of Asoka that the Parthians threw off the Seleucid yoke, and it was later still when Bactria withdrew from Greek control and made itself independent by gradual stages. Probably missionary work amongst the Greeks simply meant amongst the people of Bactria and Sogdiana, which were under Greek rule and which afterwards appear as strongholds of the Buddhist religion.

(2) Did Buddhism Spread West?

Asoka endeavoured to spread Buddhism by a series of edicts in which he set forth the " Law of Piety ". In the publication of these edicts he followed the precedent of the Achaemenid kings of Persia, who had carved decrees on the rocks at Bahistan and elsewhere. Some thirty-four edicts of Asoka are known to survive, fourteen on the rock face, seven on pillars, others in less prominent places. They are widely scattered from Afghanistan to Mysore. They were written either in the Prakrit language or in the vernacular of the locality : one is in three vernaculars, the Magadha dialect one of them. Though Prakrit is a later development from Sanskrit,[7] these are the earliest Indian documents, for the Sanskrit Vedas were transmitted orally and not committed to writing until long after the time of Asoka. The edicts are in the script known as Karoshti, a modification of the ancient Aramaic writing which had been introduced into the Punjab by the Persians

[7] See note on page 186.

in the fifth and fourth centuries B.C. The use of this means of instructing the people obviously implied that there were those who could read what was written, and this strongly suggests that Viharas or Buddhist monasteries were planted out near where the inscriptions were placed so that monks could read and enlarge upon the teaching they contained. It can hardly be supposed that a literary education, even of the most elementary sort, had spread amongst the tribes of Central Asia.

In the Bhabra edict an address to the monastic order generally, we read of the " conquest by the Law of Piety . . . won by his Sacred Majesty in his own dominions and in all the neighbouring realms as far as 6,000 leagues where the Greek king named Antiyaka (Antiochus II) dwells, and north of that Antiyaka, where dwell the four kings severally named Turamay (Ptolemy), Antigonus (Gonatus), Maga (Magas of Cyrene), and Alexander (of Epirus ?), and in the south the (realms of the) Cholas and Pandyas, with Ceylon also : and here, too, in the king's dominions, amongst the Yonas (Greeks) and Kambojas and Ptinkas, amongst the Andhras and the Pulindas, everywhere men follow his Sacred Majesty's instruction in the Law of Piety ". On the face of it this seems to claim missionary enterprise throughout the Greek world, not necessarily that the princes were converted, but that generally they received Asoka's mission graciously (Senart in *J.A.* (1885), 290 *sqq.*). Magas of Cyrene and Alexander of Epirus died about 258 B.C., so probably were not alive at the date of this decree.

Besides these inscriptions Asoka left cave temples and rock carvings. There are also early coins and tokens representing sacred objects of the Buddhist religion, the elephant of which Buddha's mother dreamed before his birth, the tree under which his enlightenment took place, the wheel which represents his teaching, and the burial mound which marked the place where he died. How far Buddhism really spread into the Greek world is problematical. A Buddhist gravestone found at Alexandria and a monument definitely Buddhist in its symbols found at Axum are the main traces, but both these places were trading ports closely connected with the Indian trade, and it would have been likely enough that an Indian merchant or traveller may have died in either place. The

Ceylon chronicles describe Asoka as having converted a large number of Yonas or Greeks, and as having sent a Yona named Dhammarakkita as a missionary to Aparanta on the coast of Gujerat. No doubt Yona simply means an Asiatic who lived under Greek rule.

According to the Puranas the Maurya dynasty of Magadha came to an end in 184, when the last king was murdered by a fanatical Brahman named Sunga Pushyamitra, who seized the throne and began to persecute the Buddhists. The result of this was that Buddhists favoured the Greek invaders whenever the Seleucids sent forces to recover the territory which once had been theirs in India.

The Ceylon Buddhist chronicle, known as the *Mahavamsa*, probably of the fourth century A.D., contains versions of some early Indian traditions, and speaks of a *thero* or Buddhist abbot of Yona (Yavana) who gathered round him 30,000 ascetes in the neighbourhood of Alasanda, the capital of the Yona country (*Mahavamsa*, trs. Turnour, p. 171). It would be absurd to suppose that Alasanda denotes Alexandria in Egypt and that there were 30,000 Buddhist monks there. The *Mahavamsa* pictures this assembly of ascetes as taking place at the foundation of the Maha thupo or " great tope " of Rusawelli by King Dutthagamini in 157 B.C., and gives details which are of a fictitious character, of stones which moved into place by themselves, of work done by demons (*dewos*), which cannot be regarded as historical. The thero or abbot was the same Dhammarakkito who is described as being the Greek Buddhist sent to preach in Gujerat. There are several Alexandrias, some in Bactria, Sogdiana, and Gandara, all lands under Greek rule until about 130 B.C., and so naturally classed by Indian writers as Yavana " the land of the Greeks ". The Alexandria intended in the Mahavamsa may have been Alexandria " under the Caucasus ", the " Queen of the Mountains " of the Alexander romance. It was in Opiane, and Alexander founded it on his way northward by the road from Seistan to Kabul as he went towards the Hindu Kush " in radicibus montis " (Curtius, vii, 3, 23). Tarn shows good reason for believing that this Alexandria and Kapisa formed a double city, such as was not uncommon in Asia, and the Greek half, Alexandria proper, was on the west bank of the River Panjshir-Ghorband. Its exact site is not known as the locality has not yet been excavated. This was an area in which Buddhism spread in the age of Asoka and it long remained predominantly Buddhist. There are great Buddhist sculptures at Bamyan close by.

The chief argument against Buddhist activity in the Greek world is the very defective knowledge displayed of anything that can be recognized as Buddhist in extant remains of Greek and Roman writers save in those few who, like Megasthenes, had visited India or had met Indian envoys who came to western lands. Megasthenes was the Seleucid agent at the court of Magadha from 301 to 297 B.C., but his work on India is known only in citations by Strabo and Clement of Alexandria. Strabo mentions Indian priests known as Σαρμάνας, which probably represents the Buddhist *Sramanas* (Strabo, xv, 1, 59). Clement of Alexandria refers to the Σαρμαναῖοι Βάκτρων undoubtedly Buddhist priests or ascetes of the Bactrians, and to two classes of gymnosophists known as Σαρμᾶναι and Βραχμᾶναι (Clemens Alexandrinus, *Stromat.*, i, 15). In this he is citing Megasthenes. The latter term doubtless means Brahmans, whilst the former seems to represent Buddhist *Sramanas*. From some unknown authority he quotes that " there are some of the Indians who trusting in the precepts of Buddha (Βοῦττα) because of his exceeding holiness regard him as (εἰς for ὡς) a god " (ibid.). But he misses the identification of these worshippers of Buddha with the Σαρμαναῖοι or Σαρμᾶναι already mentioned. Else-where he speaks of certain Indian ascetes known as " holy men " (Σεμνοί), who are not to be classed with gymnosophists and have sacred buildings in the form of pyramids (ibid., 3, 7), and these no doubt were Buddhists. Megasthenes' remark that there are Indians who honour Buddha as a god is interesting as showing that in his days Buddhism was already passing out of its primitive stage in which Buddha was simply a religious teacher and was entering the later development in which he was deified. The deification of Buddha is usually ascribed to the spread of the principle of *brakti* or personal devotion to a deity, a principle evolved in the Bragavata religion which penetrated Buddhism about 100 B.C. and led to the representation of Buddha in human form, the early images strongly influenced by Greek art, especially in the details of their drapery.

An account of Buddhism was given by the Syrian writer Bar Daisan, who obtained his information from Indian envoys passing through Syria on their way to Elagabalus or some other Antonine emperor. He does not refer to Buddhists by name, but speaks of Σαρμαναῖοι : this is cited by Porphyry (*De abstin.*, iv, 17) and by Stobaeus (*Eccles.*, iii, 56, 141).

In the embassy sent by a king of Pandya to Augustus somewhere about A.D. 13, there was an Indian fanatic who burned himself alive in Athens, an event which made a great stir. The incident is described by Nicolaus of Damascus, who met the embassy at Antioch and his account is quoted by Strabo (xvi, 1, 73, 270) and by Dio Cassius (liv, 9). This fanatic's tomb was still to be seen in the days of Plutarch and bore the inscription—

ΖΑΡΜΑΝΟΧΗΓΑΣ . ΙΝΔΟΣ . ΑΠΟ . ΒΑΡΓΟΣΗΣ.

The first word possibly represents *Sramanokarja* or " teacher of ascetes ", which denotes one of the superior class of Buddhist clergy. Probably the name ΒΑΡΓΟΣΗΣ means Barygaza on the Indian coast.

This rather scanty and scattered information represents what could be learned from Indian embassies coming to the Roman Empire or from travellers' reports. It gives no indication of anything which would have been gained from Buddhist propaganda in the Graeco-Roman world and this, in conjunction with the silence of the Ceylon chronicles, seems conclusive. The belief that there must have been effective Buddhist missions as far as Egypt rests on the assumption that the Christian ascetic life which arose in Egypt necessarily had a Buddhist origin, but this is not proved : Egyptian monasticism had an independent origin which can be satisfactorily traced. The later philosophical schools of Alexandria were fond of referring to Indian ascetes, but do not show any real familiarity with them. There remains the possibility that the teaching of the Gnostic sects which arose in Mesopotamia give signs of Buddhist influence. That seems likely, but here again there is as yet no definite proof.

(3) BUDDHIST BACTRIA

About A.D. 45 the Romans obtained greater familiarity with the phenomenon of the monsoons and as a result there was a quickening of the intercourse between the western world and the coast of India, and especially with North-West India, where at the time was the well ordered and prosperous state of Kushan. This made the Kushan ports marts for trade with the Roman Empire and through them great wealth passed into the Indian world. India also benefited culturally from this intercourse with the west, as appears from the impress of Greek thought on Indian philosophy. The rules of the syllogism in logic, as given by Carake-samhita (*circ.* A.D. 78) and Aksopada (*circ.* A.D. 150) are entirely drawn from Aristotle (cf. M. M. Satis Chandra Vidyabhusana in *JRAS.* (1918), 469).

Kushan was a wealthy and prosperous state when its third king, Kanishka, ascended the throne in A.D. 123. A great warrior he had conquered Kashmir and set up his capital at Purushapura (Peshawar). He was a convert to the Buddhist religion and used every opportunity to spread its teaching through his kingdom, which spread over a great part of North-West India. Under Kushan rule Balkh or Bactria came to be known as " the little Rajagriha ", second in sanctity only to the area where Buddha had actually lived and taught. Buddha had never lived in Balkh, but the country possessed an exceptional number of *topes* or shrines containing some portions of his body or fragments of his clothing. Many of those shrines owed their erection to King Asoka, and in their design show plain traces of Greek art. At Kanishka's court were many sculptors who had been trained in the frontier state of Gandhara, where Greek models still dominated local art, and that Gandhara-Greek art spread through Chinese Turkestan, then into China, and ultimately to Japan, carrying with it a form of sculpture and decoration which clearly shows its Greek origin (cf. A. Foucher, *Beginnings of Buddhist Art*, trans. F. W. Thomas, 1917).

It is said that Kanishka, in his enthusiasm for Buddhism, carried off the Buddhist saint Asvaghosa to his capital. This holy man was a convert from Hinduism and joined the Buddhist sect, or rather school, of the Sarvastivada, whose teaching was mainly based on the doctrine of saving grace by faith. Under Kanishka the Buddhists held another general council which resulted in the composition or revision of the authorized commentaries on the three sacred Pitakas. From the

Sarvastivada sect arose the Mahyana doctrine which gradually replaced the older Buddhist doctrine called Hinyana, Buddhism like other religions passing through a series of developments. The Buddhist aim was the path of deliverance from this world of illusion. The vehicle or *yana* in the older teaching was asceticism by which man might with difficulty approach the Buddha : this the reformers called *hinyana* " the lesser vehicle ", their own teaching was that by faith a man can enter into union with Buddha, and this they called *mahayana* " the greater vehicle ".

Although the revival of the Hindu religion gradually led to the extinction of Buddhism in India, that religion long remained a means of promoting international intercourse, being free from the caste restrictions of Brahmanism. Balkh had become Buddhist under its Kushan rulers and was visited by foreign pilgrims, especially from China and Ceylon. About 405–410 the Chinese Buddhist Fa-hien travelled to Northern India in search of authentic texts of the Buddhist monastic books and has left us an account of his travels. He says that between the Indus and Jumna there was a series of Buddhist monasteries and thousands of monks. This was under the Gupta King Chandragupta II. Fa-hien states that the people of Khotan were all Buddhists, mostly of the Mahayana school. In Pataliputra there were two monasteries, one of the Hinyana school, the other of the Mahayana.

After Fa-hien there was fairly regular intercourse between China and Northern India and Balkh, Chinese pilgrims visiting lands so rich in relics of Buddha. This did not continue quite until the Muslim penetration of Persia, for before that event it seems that there was a revival of the Mazdean religion in Persia, and some at least of the Buddhist monasteries in Balkh were transferred from the Buddhists to the followers of Zoroaster.

After the sixth century, during which the Gupta dynasty was involved in obscurity, the centre of interest shifts to Thanesar, north of Delhi, where a raja named Harsha (606–646–7), after a series of wars lasting thirty-five years, produced a strong and well-ordered state. Educated by Brahmans and by Buddhist monks this monarch, at first a disciple of the Hinyana, then of the Mahayana school, evolved an eclectic type of Buddhism which he propagated with great ardour. At that time Buddhism was losing its hold on the Gangetic plain which was its original home, but it was still powerful in India, though the religion of a minority. Harsha's capital was Kanauj. Chinese pilgrims still came to Magadha and Balkh, amongst them Hiuen-Tsang, who sought authentic copies of the Buddhist scriptures and boasts of having taken home to China 150 relics of Buddha's body or clothing. He has left a description of his journeys and of the lands through which he passed, his interest mainly centred in matters connected with the Buddhist religion. Balkh he calls Po-ho, there he was well received by the governor, who told him that the land " is called the little Rajagriha, its sacred relics are exceedingly numerous " (St. Julien, *Hist. de la Vie* . . ., 64). On the west of the capital city was the great convent of Nawbahar (Skr. *nava pihara*, " new monastery "). The hereditary abbot of this monastery bore the title of Barmak, and from these Barmaks was descended the Barmakid family which became so prominent under the early 'Abbasids. In Muslim times it was supposed that the monastery of Nawbahar had been Mazdean, but Ibn al-Faqih (edit. De Goeje, 322) describes its great temple as devoted to idols and frequented by pilgrims from India, Kabul, and China. If it had been a Mazdean temple there would have been no idols, nor would there have been pilgrims from lands where fire-worship was unknown : in any case the accounts left by Chinese visitors put its Buddhist character beyond dispute. No doubt it was converted into a fire temple during the Mazdean revival which preceded the Muslim conquest. Tradition associated Khurasan with the rise of the religion of Zoroaster in Achaemenid times, and it is quite possible that Mazdeanism was inclined to treat Bactria and Sogdiana as sacred from that association.

Another distinguished Chinese traveller was I-tsing, who made his pilgrimage during A.D. 671–695, and for about eleven years (675–685) was an inmate of the Nalanda monastery. As Buddhism lost its hold on India it took more and more an international character and assumed importance as supplying the motive for steady intercourse between the Far East and Central Asia, connecting China with Magadha and Balkh in religious interests and so ultimately with the Hellenic world. In tracing the part played by Buddhism no

attention has been paid to Tibet, although Buddhism is said
to have been introduced there by King Srong-Ban Gampo,
the founder of Llhasa, in 629–650, for Tibetan Buddhism
really traces from monks of Magadha who conducted missionary
work in Tibet as late as the eleventh century.

In connection with the strongly marked Buddhist element in
Eastern Persia reference should be made to Bamiyan, the chief
city of East Ghur, south of Balkh, where was an important
Buddhist centre. In the thirteenth century Yaqut described
two great images of Buddha there in a large chamber excavated
in the mountain side, images known as *Sushk Bud* " the red
Buddha " and *Khing Bud* " the grey Buddha ", which still
existed in his days. They are mentioned also by Qazwinu.
Bamiyan was destroyed by Changiz Khan.

It seems fairly certain that Buddhism promoted intercourse
between the Graeco-Roman world, especially Alexandria, and
the parts of India comprised in the Gupta Empire, more
particularly at Pataliputra, where Indian scholarship shows
distinct traces of Greek influence.

(4) IBRAHIM IBN ADHAM

There is a curious addendum to the history of Buddhist
influence on Islam in the life of the saint Abu Ishaq Ibrahim
ibn Adham, who died between 776 and 783. This saint was
a noted ascete, a type not very common in primitive Islam.
He perished in the course of a naval expedition against
Constantinople, which may be taken as an historical fact.
Less convincing, however, are the details of his earlier life.
It is related that he was a prince of Balkh (Bactria) who was
converted to the service of God whilst engaged in hunting
and forthwith abandoned all worldly honours and material
possessions in response to the Divine Call. But careful examination of his biography shows that it is a Muslim version of the
life of Gautama Buddha, and it seems reasonable to suppose
that this came into Muslim hands through Marw, where
there was a strong Buddhist tradition. Possibly the story was
introduced into Muslim circles during the earlier 'Abbasid
period.

CHAPTER X

THE KHALIFATE OF DAMASCUS

(1) CONQUEST OF SYRIA

A MAP of the physical features of Western Asia and North-East Africa shows two important river valleys, one of the Tigris and Euphrates, the other of the Nile, and between them high ground, broken rather abruptly by the Red Sea. These conditions are due to geological changes with which we are not at present directly concerned : we start from a point when the two great river valleys already existed, with a good deal of barren highland between. Those two valleys were the homes of two primitive civilizations, which was the earlier is still not decided. In both cases the rivers concerned overflowed and flooded the surrounding country regularly every year, and the particular river-valley culture which grew up there was based on the artificial control of these regular inundations, draining the swamps and directing the water so as to fertilize the fields. It is commonly assumed that in primitive society land was held in common, each member of the tribe entitled to his share, but not to permanent ownership of any particular piece. Whether this is universally true is disputed, probably it does apply so long as tribes are nomadic. But in the river-valley culture of Mesopotamia and Egypt the productivity of each field depended a great deal on human labour, irrigating and draining the land, so that private ownership developed at a fairly early date and population became stationary. The people of the barren highlands between the river valleys remained nomads, not recognizing the rights of private property and in all respects at a much more primitive stage of social evolution than the settled inhabitants of the valleys. The life of those nomads was hard and bare, it generally was, and still is, on the border of starvation ; there always was a temptation for those nomads to raid the fertile and productive settlements, and when their numbers became too great to be able to make a living out of the meagre resources of the desert highlands, they tended to overflow into the valleys. Thus all through ancient history the kingdoms of Assyria, Babylonia, and Egypt found their nomadic neighbours a perpetual nuisance, and it was always necessary to provide for the protection of the frontiers, those frontiers being the precise level at which it ceased to be practicable to raise the water from the rivers to irrigate and fertilize the land. Whenever military power so far decreased as to make the guardianship of the frontiers insufficient to protect the settled country from Arab raiders, then Arabs came down to raid the country, then to settle in the rich and productive territory and reap the benefits of a cultivation at others' expense, usually subjugating and sometimes enslaving the unwarlike population already settled there.

Such a raiding and settlement took place towards the end of the seventh century A.D., when the raiding Arabs were united in a religious fraternity based on the religion taught by the Prophet Muhammad. It does not seem that Muhammad himself had any project of foreign conquest, but such conquest followed because the people of the area invaded were exhausted by prolonged warfare, distracted by internal divisions, and disaffected by harsh government, though some of that harshness was the inevitable result of war conditions. The success of their expeditions seems to have surprised the Arabs and encouraged them to undertake the permanent occupation of the countries they had conquered. They had not the least desire to cultivate the soil or settle down to agricultural work, their idea was to establish a military occupation and live on the fruits of the toil of the native inhabitants. In this they were, no doubt, influenced by the precedent of the Arabs stationed along the Persian and Roman frontiers. On both those frontiers it had been found impossible to dislodge the Arab tribes and both countries tried the same solution, permitting the tribesmen to settle there and paying them a subsidy on condition that they guarded the frontier against any other Arabs who tried to invade the Persian or Roman territories. The Arabs already settled and paid were greatly envied by the hungry nomads of the desert, their existence seemed an ideal one, and when they conquered the eastern provinces of the Roman Empire and the kingdom of Persia they counted on living a similar kind of life, occupied in hunting and occasional warfare and supported by the tribute paid them by the conquered population. Nor were the conquered people unwilling to toil and pay tribute, as they were to be disarmed and freed from the hated military service which was the task they most disliked.

It is a debated point whether Muhammad intended his religion to be a universal one, or for the Arabs alone. Qur'an 34, 27, says, " We have sent thee to mankind at large, to announce and threaten." But the context shows that this refers to the Prophet warning men of the approaching end of the world and is itself one of the signs that the end is near, and is thus interpreted by tradition (Bukhari, Ṣaḥiḥ, i, 93, 4a, d. 1 : Muslim, Ṣaḥiḥ, i, 53, 55). It is necessary for all Arabs to believe in Muhammad if they are to escape hell (Muslim, i, 54), but it is not stated to be necessary for non-Arabs to believe, though those who join gods to God, that is to say, are polytheists, are doomed to hell in any case. As regards the non-Arab world, the Qur'an seems to contemplate conquest rather than conversion (Qur., ix, 19–23). One passage in the Qur'an says, " and one day We will summon up in every people a witness against them from themselves : and We will bring thee (Muhammad) up as a witness against them : for to thee We have sent down the book which makes all things clear, a guidance and a mercy, and glad tidings to those who reconcile themselves with God " (Qur., 18, 91). In another place the Qur'an says, " thus We have made you a central people that ye may be witnesses in regard to mankind, and that the Apostle (Muhammad) may be a witness in regard to you " (Qur., 2, 137). But these passages fall far short of a definite missionary command to go forth and preach Islam to all the nations of the earth.

In the later years of his ministry Muhammad preached his religion to all the Arabs and endeavoured to unite the tribes in one confederacy. " Fight until there is no more civil discord and no worship save that of God " (Qur., 2, 189) : " fight against those who oppose you, but do not attack first " (Qur., 2, 186), " kill and expel them " (Qur., 2, 187) : " when the sacred month is over slay the polytheists, but spare the pagan Arabs who are in league with you " (Qur., 9, 1–4), but these commands were preparatory to the reduction and unification of Arabia. They find their best explanation in Muhammad's own conduct, for he strove hard to draw all the Arabs into his fold, though tolerating those who were

" people of the book ", i.e. Christians or Jews. His attitude
was endorsed by the policy of the early khalifs, men who had
been his intimate companions and trained by him, men who
knew his outlook as no others could, and they for some time
insisted on all converts to the religion of Islam also becoming
members of an Arab tribe. Much weight must be attached to
the expressed reluctance of the older Muslims to spread wider
into the outside world lest the multitude of strangers brought
in as converts might outnumber the native Arabs, by their
influence changing the character of their religion and mode of
life, apprehensions which subsequent events showed were
justified.

The traditional and legendary biography of Muhammad
attributed to Ibn Ishaq and known to us in an edition
expurgated by Ibn Hisham represents him as sending letters
to foreign monarchs, the King of Persia, the Roman Emperor,
and others, inviting them to become Muslims, but that
biography was composed in its earliest form about a century
after Muhammad and contains a great deal which cannot
be regarded as historical.

There can be no question that Muhammad intended to
include all the Arabs in the brotherhood of Islam. Those
Arabs were the inhabitants of Arabia, not quite the artificial
Arabia marked on the atlas, but all the desert highlands of
Western Asia, spreading up into a tongue in Syria. In that
northern area, between the two great monarchies of Parthia-
Persia and Rome were the two groups of border tribes subsidized by the monarchies and to some extent settled and
civilized. Muhammad was very anxious to draw these border
tribes into his fraternity. The Arabs along the Persian frontier
had some grievances against Persia and joined the Muslims,
but threw off their allegiance as soon as Muhammad was
dead. In order to gain the Arabs of the Syrian (Roman)
frontier Muhammad sent an envoy to invite them to embrace
Islam, but that envoy was killed at Bosra, a crime against the
Arab tradition of the sacred character of an ambassador.
So an army was sent under Zayd to avenge this. But the
border Arabs being in Roman employ obtained the help of
Roman legionaries and defeated the Arabs. For some time
no further action could be taken as the Arabs were busily
engaged elsewhere, but in 632 an army was assembled and

preparations were made to invade Syria. But Muhammad died whilst the expedition was waiting to set out. Then Abu Bakr was appointed khalif or " successor " and ordered the army to set out. After forty days it returned laden with booty, so there was no difficulty in raising new forces. In 634 these forces invaded Syria, where they met small resistance, and that only from an ill-trained local militia. No one as yet supposed that the Arabs were venturing on more than an ordinary raid, nor do the Arabs themselves seem to have thought that they had undertaken more than that.

Certainly these Arabs were not fanatics who tried to force their religion on the conquered : they preferred them to remain toilers as before and themselves to live on the produce of their labour. Such was the system laid down in the " Constitution of 'Umar ", an apocryphal production of later date, but indicating in general outline what was the earlier Arab policy. The picture sometimes given of a host of fanatical Arabs rushing forward with a sword in one hand, a Qur'an in the other, and forcing people to turn Muslims or be killed is very far from fact. The cynical Arab is not inclined to be a fanatic. There have been plenty of fanatical Muslims, but they were not Arabs but converts of other races who were converted to Islam at a later date. The Arabs did not force the people they conquered to embrace their religion, they left the conquered population to follow their own religion, laws, customs, and use their own languages. They were to be tribute producing and the Arab ideal was to live at ease on the product of their labour.

In Syria, which was of primary importance because in 661 the khalif with his court and government settled in Damascus where they remained for more than eighty years, the Arabs found themselves rulers of an area which had been a Roman province subject to the fully developed Roman law and with a highly organized administration. This they took over as it was. Any Roman officials who wished to remain under Roman rule were given every facility to remove to some part which still remained Roman. Many did so remove, but many others were content to live under Arab rule, and of these numbers rose to high office and dignity in the Muslim state. For the first twenty years or more the records continued to be kept in Greek, and the civil service was almost exclusively

Christian. There already were a number of Arab tribes settled along the border, they had been subsidized by the Byzantine government as defenders of the frontier, and these were Christians. As old established settlers they had become wealthy and considered themselves socially superior, to the Muslim invaders, poor hungry nomads of the desert, and had no hesitation in asserting themselves, the Muslim Arabs admitting their claims to aristocratic status. Some of the ruling dynasty married women of these Christian tribes, and that was rather resented by the Muslims. Under the khalif 'Abd al-Malik (685–705) there was a good deal of jealousy because the Christians had a monopoly of all the posts in the civil administration, and the khalif tried to employ Arabs in their place. But the change was not successful, the Arabs did not understand the details of business and the Christian officials had to be restored. This is easy to understand because the oriental practice is, not to draw up accounts so that an outside auditor can understand and check them, but to keep them in such a way that nobody but the established officials can possibly understand them : it is done deliberately so that the established officials may keep the business in their own hands and secure a permanent monopoly. The most that 'Abd al-Malik could do was to get the records kept in Arabic instead of Greek, and to use Arabic on the coinage. Bishop Arculf of Gaul made a tour of the Holy Land about 700 and speaks with much appreciation of the hospitable way he was received by the Muslim rulers, the freedom with which he was allowed to travel about, and the generally friendly attitude of the Arabs and their rulers. Until the days of the Crusades Syria and Egypt were practically Christian lands under the rule of the Muslim Arabs, their rule mainly confined to the collection of taxes, and that they did very thoroughly.

In the earlier period of the 'Umayyad khalifate at Damascus there was even a fashionable tendency to deride Islamic ways and customs. This is well illustrated by the poetry of Abu Malik Ghiyath ibn Salt ibn Tariqa al-Akhtal, who was born at Hira about 640 and died about 710. He belonged to the Taghlib clan of the Jusham ibn Bakr tribe and lived and died a Monophysite Christian. His poems refer to St. Sergius, the Holy Cross, to monks, and he uses Christian oaths, though there are very few direct references to Christianity in his Diwan. He refused to change his religion (Diwan, p. 154), and derided those whom he described as becoming Muslims by pressure of hunger rather than by conviction (ibid., 315). He composed poems in honour of Yazid, the son of the khalif Mu'awiya, his brother 'Abdallah, and others of the royal family. He was formally recognized as poet laureate by 'Abd al-Malik, whom he celebrated as well as his relations and derided their enemies, a real courtier. In his poems there appears evidence of the survival of ancient pagan Arab usages in the days of the 'Umayyads, and some striking instances of the tolerant attitude of that dynasty. Many of his verses contain biting sarcasms on Islam, and such passages have prevented many Muslims from full appreciation of his poetic merits, but in his day he and his rival Jarir were the leading poets of the Arabs. He particularly expresses his contempt for all those who abandoned their ancestral religion, Christian or pagan, to conform with that of the reigning monarch. The most admired passage in his works is his panegyric of the 'Umayyads (Diwan, 98–112). In spite of his contemptuous attitude towards Islam this poet was patronized by the khalif 'Abd al-Malik, though not greatly favoured by his successor Walid I. He probably died before the end of Walid's reign, though Ibn 'Abd Rabbihi prolongs his life to the reign of 'Umar II. Probably his death should be dated about 710.

A loose tone about religion prevailed at the 'Umayyad court, which did not find favour with the stricter Muslims, and was one of the causes of the anti-'Umayyad feeling which grew in intensity until it led to the downfall of the dynasty. The old tribal rivalries of pre-Muslim days still influenced the Arabs, and there was a deep-rooted antagonism between the worldly tone of Damascus, and the cities of Mecca and Medina, and the more orthodox attitude of those who regarded themselves as Muslims in the first place, and Arabs only in a secondary place. The only exception to this in the 'Umayyad khalifs was Walid I (705–715), who was a really religious man and put the interests of Islam before political or racial considerations. At the other extreme Yazid I (680–683) is still cursed by the orthodox as an enemy of religion. It was an army sent by him which engaged in the battle of Kerbela (10th October, 680), and was responsible for the tragic death of al-Husayn, the surviving son of 'Ali the Prophet's son-in-law. And it was an army sent by him which besieged the holy city of Mecca and (accidentally) burned the sanctuary of the Ka'ba (November, 683).

(2) THE FAMILY OF SERGIUS

Damascus, the official capital of Syria, was a partly Greek city, not so thoroughly Hellenized as Antioch. It was the seat of Christian bishops who ranked next after the patriarchs of Antioch in the ecclesiastical hierarchy of Syria : it possessed a school which, though not equal to those of Alexandria and Antioch, yet had attained considerable eminence by the time of the Arab conquest, and retained its good repute after that event. Amongst its alumni were the theologian Sophronius, who became bishop of Jerusalem (634–8), and Andrew of Crete (*circ.* 650–720), who studied there after the Arab conquest, became a monk in Jerusalem, and finally bishop of Crete. The Arabic historians say that at the time of the conquest the financial agent of the Roman government in the city was Sergius (Sarjun) who was responsible for making terms with the invaders, on which account Eutychius calls him a traitor. But the citizens, deserted by the government, had no choice in the matter and it is probable that everyone supposed that the Arab attack was no more than a raid on a large scale and that after plundering the town the Arabs would go back again to the desert. The governor of such a city normally was a financial agent whose duty it was to raise the imperial taxes and commonly bore the honorary title of Patricius which had been granted to all superior officials by Constantine. He had been appointed by the Emperor Heraclius, but like many other officials continued in office after the Arab conquest under Mu'awiya, when he was governor of the province, and remained when Mu'awiya became khalif. Finally he acted as minister of finance for the whole Islamic state and paymaster-in-chief of the Arab army. Yet he remained a Christian, and long after becoming minister of finance built a Christian church. His son was treasurer under 'Abd al-Malik, and his grandson was chief minister under some of the later khalifs. The office and title of wazir had not yet come into existence.

It is said that the second member of this family purchased a slave named Cosmas, a monk who had been captured by the Arabs during a raid on Italy, and employed him as tutor to his son John. When Cosmas had taught him all that he could he begged permission to retire to a monastery, and on obtaining leave he went to the Laura of St. Sabas, near Jerusalem. The author of this John's biography was John of Jerusalem who lived in the tenth century, a good while after the events he records, and like many hagiographers of the time used freely matter which would now be regarded as legendary, but the main lines of John's life seem to be reliable. It appears that this John was the son of Sergius, afterwards known as St. John of Damascus, son of an important official in the Arab state, was himself attached to the court and acted as " chief adviser " to the khalif, probably Hisham (724–743). After serving the khalif for some years John asked leave to resign, and followed his tutor to the Laura of St. Sabas where, after a period of rigorous discipline, he was ordained to the priesthood some time before 735. He died before 743. To him is due the earliest treatise on the controversy between Christianity and Islam, the " Disputatio Christiani et Saraceni " which is printed in Migne's *Patrologia Graeca*, xcvi, 1335–1363. This work shows that there was great freedom of religious discussion in eighth-century Damascus, and that Christians were permitted to criticize the established religion very freely. The text says, " When the Saracen says. . . . You reply. . . ." John gives proof of a good knowledge of the Qur'an and familiarity with Muslim ritual and doctrine. The identification of St. John of Damascus with the son of Sarjun ibn Mansur was first made by William of Tripoli.

Theodorus Abucara (d. 826) was St. John's pupil, and he also left works on the controversy with Islam. Obviously there was unrestrained intercourse between the two religions and no reluctance was felt about discussing religious differences quite freely. It may reasonably be supposed that such intercourse introduced the Muslims of Damascus to a general knowledge of Christian theology and philosophy, and within the next following generations ideas and problems suggested by Greek philosophy appear leavening Muslim thought.

A parallel infiltration of Greek thought took place in jurisprudence so that the earliest speculations of the Muslim jurists are tinctured by theories gathered from the Roman law which itself contains elements gathered from Stoic philosophy, and thus Greek philosophical teaching was passed on to the Arabs through a legal medium. Roman law at the time of the Arab conquest circulated in the eastern provinces in Greek, and slightly modified by local conditions, but it contained the Stoic principles which the lawyers of Rome had drawn from Greek sources. Prominent among these philosophical-legal theories was the doctrine that man has an innate sense of what is just and right, of what the Stoics called the Law of Nature. This was also assumed by the early Muslim jurists who appealed to " opinion " to supplement and even to supplant the written law when cases arose for which no provision had been made. Here, however, it is to be noted that the earlier indications of this Stoic doctrine appear, not in Syria where the Roman law was established, but in 'Iraq, and especially at Basra. That the Arabs were first brought into contact with Roman law in Syria and Egypt is certain. They had conquered those provinces and found there a complicated system of land tenure, contractual law, and commercial legislation dealing with things of which the simple nomads of the desert had no previous knowledge. Much of this they adopted, indeed such adoption was inevitable, and it henceforth was incorporated in Muslim law. It is true that there are some branches of law which had already been incorporated in Jewish law, and those may have come through a Jewish medium to the Arabs, but it is more probable that most of the law dealing with land tenure, contract, usufruct, inheritance, and certain other matters came direct from the customary law already prevalent in Syria and Egypt when the Arabs conquered those lands, and that established law which they found there was the Roman law.

In the parallel case of theology it may be noted. (1) One of the earliest theological problems faced by the Muslims was that of the eternity of the Qur'an. The older doctrine was that it was eternal, co-eternal with God. Then the problem arose, if this were so, then God is not the one source and creator of all things, for there must have been an uncreated Qur'an, like a second god, side by side with the One. This was hotly debated. The sect of the Mu'tazilites held that the Qur'an was created by God and, as the author must precede the work produced, the Qur'an must be less eternal than God.

The orthodox maintained that the Qur'an is co-eternal with God, though the words in which it is expressed, like the paper on which it is written, may be created and so not eternal with God. Ultimately the orthodox opinion prevailed and the Mu'tazilites became extinct, for those who now call themselves by that name in India are modernists of recent date, in no way connected with the old Mu'tazilites. The point is that in the discussions between the Mu'tazilites and those who adhered to the orthodox theory very much the same arguments are used as were employed in the Arian controversy in the Christian Church, much of this repeated in the writings of St. John of Damascus. In Christian theology the term " Word " was used as a mystical name for Christ, as it was used by St. John in the fourth gospel, whilst the Muslims used the same term to denote the written word in the Qur'an, but in general the arguments are the same. It is difficult to avoid the conclusion that the problem involved was suggested to the Muslims by Christian theology, the teaching of St. John of Damascus, or some other.

(2) Another early problem concerned the freedom of the will. If God is almighty, then everything is overruled and directed by him. Therefore man has no freedom. But Greek ethics assumes that man is responsible only when he has free choice, and the Qur'an gives commands and prohibitions in such a way as to imply that man has such a choice. The Mu'tazilites argued that as God is just, he will only punish men when they have been free to choose and have chosen wrong. From this and the preceding point the Mu'tazilites called themselves " the People of Unity and Justice ", of unity because admitting only One Creator, One Source, and so asserting that the Qur'an is created, and of justice as defending the freedom of the will as necessary for man's responsibility.

(3) A third problem concerns the qualities of God. God as the sole source of all that is must be a unity, not compounded : so God has no qualities or accidents, he is himself essence. The only attributes that can be predicated of God are negative, that he is eternal or having no beginning or end, that he is infinite as having no limitations, and so on. This, however, seems to be contrary to the Qur'an which does apply qualificative adjectives to God. The orthodox opinion is that these attributes given in the Qur'an may be applied to God because

they are so applied, but they do not convey the same meaning as they would if applied to men, nor do we know what they imply. This was already taught by Plotinus and other neo-Platonists, and it would seem that the problem and its solution was borrowed by the Arabs from them.

At first sight it seems that these traces of Greek influence on Arab thought most likely connect with Syria where Arabs and Christians had very free intercourse ; but the first traces of that influence appear in Mesopotamia towards the middle of the eighth century. Greek influence may have been applied at more than one point, or may have spread from one area to another. It must be admitted that we have very little evidence of philosophical or theological speculation in Syria under the 'Umayyad dynasty, the dynasty which began with Mu'awiya : such matters seem to have made little appeal to Arab interest at that period. The beginnings of speculative thought in philosophy and theology and of interest in scientific research arose in Mesopotamia, and more especially in Basra, to a less degree in Kufa. These two cities were in the area where were the ancient cities of Hira and Jundi-Shapur, and it is quite possible that a general influence due to intercourse between Muslims and Christians had been engendered before the direct transmission of Greek science from Jundi-Shapur to the Muslim community had commenced.

(3) THE CAMP CITIES

After their first outspread and contact with the Roman and Persian armies, the Arabs set themselves to learn the methods of warfare used by the Romans, realizing that something different was now required from the rapid raids and retreats which had sufficed for desert warfare. The Byzantine writer, the Emperor Leo Tacticus, describes the Arabs as imitating the order and discipline of the Roman army in all details. And that was natural, for the most influential Arabs under the 'Umayyads were those of the Syrian border who had been trained as auxiliary Roman forces. At the same time it must be admitted that the Persians also had already endeavoured to copy Roman military methods. One of the new forms of warfare was the use of engineering for besieging fortified cities and for constructing fortifications for their own defence.

For this latter purpose they imitated the rectangular fortified
camp characteristic of Roman military methods.   In each
conquered area they planted such camp cities, often on ill-
chosen sites. In Palestine the chief such camp city was Jabia,
in Egypt it was Fustat, in Ifriqiya, Qairawan.   But none of
these were of so great importance as the two camp cities in
'Iraq, Basra founded by 'Utba ibn 'Azwan in 635 or 637, and
Kufa founded by Sa'd ibn Waqqas a little later. These played
a very leading part in the history of Islam.

When the 'Umayyads seemed to be secularized and
indifferent to religion, and their laxity spread, as it did, to
Medina and Mecca, many of the stricter Muslims were greatly
discouraged and removed from those places such as Medina
and went out to one or other of the 'Iraqian camp cities,
which thereby became the homes of orthodoxy and incidentally
of resistance to a khalifate commonly regarded as disloyal to
religion.

The intellectual life and interests of Basra and Kufa were
directed by religion and centred in Qur'an study and theological
sciences more or less connected with the Qur'an. At first these
sciences were chiefly those concerned with the Qur'an text
and that especially meant grammar and lexicography, but
later on opened out so as to include jurisprudence, tradition,
and philosophy, all to a great extent directed and tinctured
by ideas gained from Greek studies. Greek authorities were
not used or read, but there are clear indications that their
substance had filtered through and at Basra and Kufa impinged
on Arabic culture far more than was the case in Damascus.
It must not be overlooked that Hira, the great Nestorian
stronghold, was not far from Basra and a good deal of its
population drifted to the camp city.

Grammatical and literary studies began at Basra with Abu
l-Aswad ad-Du'ali, the friend and confidant of the Prophet's
son-in-law 'Ali.   It naturally happened that many of the
people of 'Iraq who had learned Arabic only late in life when
they were converted to Islam committed many solecisms in
reading the text of the Qur'an, and these errors distressed 'Ali.
So he appealed to ad-Du'ali to draw up some rules for the
guidance of those who were not well used to the use of the
only language permitted for prayer and reading the revealed
word. But ad-Du'ali was prevented from carrying out this

command by 'Ali's murder on 21st January, 661, and he was reluctant to take any steps to assist the governor Ziyad ibn Abihi whom he regarded with disapproval because he, after serving 'Ali, had transferred his services to the 'Umayyad usurper Mu'awiya. Though Ziyad renewed 'Ali's request ad-Du'ali held back and did nothing. Then one day he heard a reader mispronounce two vowels in the text of Qur., 9, 3, so as to pervert the sense from " God is free from (the covenant of) the idolaters, and His Apostle (also is free) " into " God is free from (the covenant of) the idolaters and (from the covenant of) His Apostle ", and this misrepresentation of the inspired word so shocked him that he forthwith began to devise methods to prevent similar errors. For this purpose he introduced vowel points into the hitherto unpointed Arabic text and began giving instruction in the grammar and vocabulary of the Arabic language. Incidentally in doing this he seems to have been to some extent influenced by Aristotle's logic, not by any of the Greek grammarians.

From Abu l-Aswad ad-Du'ali came a regular succession of grammatical students and teachers in Basra. Nearly a century later similar grammatical lectures were commenced at Kufa by Abu Muslim Mu'adh ibn Muslim al-Harra (d. 723 or 727), who at one time was tutor to the sons of the khalif 'Abd al-Malik. These two centres developed rival schools which agreed in theory, but differed in practice. As yet the works of the ancient poets, valuable in illustrating and explaining the older usages of the language, were not collected in written Diwans, but transmitted by word of mouth, often altered and inter-polated in their transmission. Aware of this the Basra school carefully criticized the poetry heard and rejected that which did not fit in with accepted standards, whilst the Kufans accepted all that was heard and are said to have used a good deal of forged material. At first sight it seems that the Basri method was better, but against that it must be noted that by that method the examples were made to fit the rules drawn up, whilst the Kufi grammarians had to adapt their rules to meet the spoken use, which is sounder.

The line of oral transmission of the two schools formed a a grammatical pedigree which led down to the great Basri grammarian Abu l-Hasan (or Bishr) 'Amr ibn 'Uthman al-Harithi, commonly known as Sibawaih (d. between 783 and 816) who, it must be noted, was not an Arab himself but a Persian and compiled his grammar under the early 'Abbasids.

At Basra arose the first indications of Mu'tazilite thought, with evidence of the solvent effect of Greek philosophical speculation on Arab theology, and in 'Iraq round about Basra were the first traces of juristic theory showing evident traces of Roman law and the philosophical theories adopted by Roman lawyers. Obviously the results of Greek influence began to appear, not in Syria where the ruling Muslims were in such close contact with Christian theology and its philosophical speculation, but in Basra, though we have no direct evidence of intercourse with Greek and Christian elements there. Damascus and its court were given over to sport and politics, and theological speculation could not have sunk very deep. Basra, on the other hand, kept alive a scholarly tradition and must have been impressed by Greek teaching, possibly through Hira, more probably through Jundi-Shapur, and so shows the first traces of Arab Hellenization.

CHAPTER XI

THE KHALIFATE OF BAGHDAD

(1) THE 'ABBASID REVOLUTION

MU'AWIYA had assumed the khalifate at Jerusalem in 661, but at once removed to Damascus, where he had already spent several years as Governor of Syria. At his accession began the rule of what is known as the 'Umayyad dynasty, which ruled Islam until 749. That dynasty suffered a break in 684 when it passed from one family to another, but the new family, descended from Marwan, was a branch of the 'Umayyad clan, so the monarchy remained in 'Umayyad hands, and that was the case until 744, when a second Marwan, not of 'Umayyad blood, assumed power by military force. The court and administration were settled at Damascus until 724 when the khalif Hisham removed to a country residence, and after that the khalifs went to Damascus only to be installed, and then retired to reside elsewhere, but the administration remained at the Syrian capital until the accession of Marwan II in 744. The court necessarily accompanied the khalif, but in 744 not only the court but also the administration were removed to Harran, which thus became the capital, and Damascus sank to the level of a provincial town, a change greatly resented by the Arabs of Syria.

Under the 'Umayyad dynasty the khalifate was a purely Arab state. Its intellectual output consisted entirely of poetry, largely of the old desert type, some of it so far modified as to reflect the tone of the courts of Hira and of the B. Ghassan, all in the spirit of the *Jihiliya* or " times of ignorance " before the coming of Islam. Its poets praised their patrons, derided their rivals and enemies, pictured the perils of the desert life, or sang the echos of ancient tribal wars. The culture and science of the Greek world found no place in their compositions, apparently meant nothing to them.

Under Marwan II the Syrian army was disaffected, the Kharijites of 'Iraq revolted and entrenched themselves in Mosul. Marwan was unable to march against them, his hold on Syria was too insecure and he had to send an army down into Arabia where there was another Kharijite revolt.

His more serious trouble, however, threatened from Khurasan in East Persia. The Persians were dissatisfied : they felt that the Arab conquest of Persia had been due to a series of accidents, to domestic revolution which undermined their military organization and to the rash conduct of their youthful king. They longed for an opportunity to try issue again with those whom they regarded as half-civilized nomads. In such conditions it was inevitable for conspiracies to flourish, indeed the whole 'Umayyad period shows the community of Islam seething with dissatisfaction and ready for revolt, partly on racial grounds, resentment at the way in which the Arabs domineered over them even after they had embraced Islam, partly on religious grounds, regarding the 'Umayyads as lax in religious observance. Amongst the Persians were many adherents of the house of 'Ali, and these regarded all the khalifs, except 'Ali himself, as usurpers. They recognized the leadership only of those descended from 'Ali. The extremer 'Alids even preferred 'Ali to Muhammad himself. All these Shi'ites, as they were called, were divided amongst themselves into many sects, but all agreed in disapproving the Arabs. At length a revolutionary outbreak took form, its centre in Khurasan, but its propaganda spread by secret agents who circulated everywhere through the world of Islam, except in Spain where Muslims had their own troubles. The identity of the person who was to be set upon the throne after Marwan was deposed was kept secret until the revolution had reached a successful end, then it was disclosed that the one selected as khalif was Abu l-'Abbas of the Hashimite clan of the Quraysh tribe, the same tribe as that to which the 'Umayyads belonged. The throne merely passed from one Arab family to another.

Abu l-'Abbas was invested with the khalifate in the great mosque of Kufa on 28th November, 749, and made it his first task to exterminate the surviving 'Umayyads and their adherents, and this he did so drastically as to earn for himself the surname of *aṣ-Ṣaffaḥ* " the butcher ". Of the deposed dynasty only one young man escaped and, after incredible dangers and hardships, reached distant Spain where he became head of an independent state, and later on his descendants assumed the title of khalif in opposition to the dynasty of Abu l-'Abbas. There are stories of other 'Umayyads who found refuge in the remoter parts of Africa, but these seem to have been adherents of the dynasty, not themselves of 'Umayyad stock.

The downfall of the 'Umayyads was a definite turning-point in the history of Islam. The 'Abbasid khalifs were no less Arab than the 'Umayyads, but they had gained their throne largely by Persian help, their chief ministers were Persians more often than Arabs, the heirs of several of the earlier 'Abbasid khalifs were educated in Persian surroundings and had Persian blood as the result of intermarriages. Persian ideas and Persian interests rivalled, in many cases displaced, Arab ideas and interests, and so to a certain extent Islam became Persianized. For all that the khalifate and its subject must still be classed as Arab : they were commanded by a ruling dynasty which was Arab, they used the Arabic language, professed an Arab religion, and held in unbroken continuity from the desert men who had conquered the Near East.

(2) FOUNDATION OF BAGHDAD

At first the 'Abbasid khalifs lived at al-Anbar [8] on the Euphrates. They had no desire to go to Syria where prevailing feeling was strongly pro-'Umayyad. But the second ruler of the 'Abbasid line, Abu l-'Abbas' brother al-Mansur determined to found a new capital. After considering various sites he at length decided to build at Baghdad, a town of considerable antiquity which had been known in Babylonian times as BAG-DA-DU, a name of unknown origin. By a play upon words later Persian writers gave this name a fanciful Persian derivation and made it mean " the Garden of God ".

In making this choice he was guided by the advice of his minister, the Persian Khalid ibn Barmak, and having resolved on building he called in the services of two astrologers to lay out the foundations and select a propitious hour for setting the first stone in position. The astrologers chosen for that purpose were an-Nawbakht, who was a Persian, and Mashallah ibn Athari, a Persian Jew,[9] of Marw.

Guided by these astrologers al-Mansur laid the first stone of his new capital towards the end of the year 762, and three years later the building was sufficiently advanced for occupation to commence. Many of the inhabitants came from the neighbouring camp cities of Basra and Kufa, both hotbeds of sedition and always restless and fanatical. The presence of these new citizens helps to explain why from the outset Baghdad showed a turbulent and troublesome atmosphere. One suburb of the city known as Karkh, which had already existed as a Persian village, was given over to Persians.

Al-Mansur desired to make his capital a city whose fame should radiate through all Islam, and for this purpose he invited to it a number of distinguished scholars, Qur'an readers and preachers, grammarians and traditionalists from the two neighbouring camp cities which had already become recognized centres of Muslim scholarship, as yet restricted to Qur'anic and theological studies. Such men of learning were then beginning to form a respected middle class which later rose by court favour to high offices in the State, but was entirely distinct from the older aristocracy of the Arab tribal chieftains of noble pedigree who had dominated Islam under the 'Umayyads. The learned men of Kufa and Basra, many already famous, formed a kind of academic aristocracy which tended to act as a check on the arrogant pretensions of the hereditary nobility who had proved a source of danger in the court of Damascus and were still disaffected towards the 'Abbasid dynasty which they regarded as semi-Persian. Unfortunately al-Mansur suffered from the unprincely vice of avarice, and the rewards he offered were so moderate and were paid so grudgingly that he earned the nickname *Abu d-Dawaniq* " father of sixpences ".

In 765 al-Mansur was taken seriously ill with some gastric disorder and was advised to send for the Nestorian physician Jirjis ibn Bukhtyishu', head of the academy and hospital at Jundi-Shapur. This was the first contact of the court at Baghdad with the family of Bukhtyishu' which afterwards played an important part in the cultural education of the Arabs. Nothing is known of the Bukhtyishu' who was the father of this Jirjis, but as the name occurs several times in the course of the history of Baghdad it is convenient to classify him as Bukhtyishu' I.

Of all the East Persians who had helped the 'Abbasid

revolution and afterwards came west to share the prosperity of the new dynasty, the most distinguished belonged to the wealthy and well-born family of the Barmakids, originally of Balkh in Bactria, but afterwards settled at Marw. This family was descended from the Barmaks or hereditary abbots of the Buddhist monastery of Nawbahar in Balkh, but had conformed to the Mazdean religion some time probably not long before the Muslim conquest, and then embraced Islam. Khalid ibn Barmak, the head of the family, was minister of finance under aṣ-Ṣaffaḥ, and was made governor of Mesopotamia by al-Mansur. His son Yahya, at one time governor of Armenia, was entrusted by al-Mahdi with the education of his son who afterwards became khalif as Harun ar-Rashid, and he appointed Yahya wazir of the whole empire and entrusted him with unlimited power. In this office Yahya showed himself a wise and just administrator, and under his guidance the empire prospered. Of Yahya's three sons Fazl was governor of Khurasan, then of Egypt, and Ja'far succeeded Yahya as wazir. But the family, after being the first in wealth, power, and honour is Islam, fell from its high estate in 803 for reasons which were a mystery to contemporaries and never have been adequately explained. Yahya died in prison in 806, Ja'far in 909. Other sons seem to have been set at liberty after Yahya's death. At the accession of al-Amin in 808 all surviving members of the Barmakid family were set free and had property and honours restored to them.

The Barmakids were keenly interested in Greek science, which was then the subject of much attention at Marw, and brought with them this taste, finding a kindred spirit already existing in the Nestorian academy of Jundi-Shapur.

Jirjis ibn Bukhtyishu', who had come from Jundi-Shapur to attend al-Mansur, remained in Baghdad as court physician until advancing years caused him to ask to be released and he retired full of honours to Jundi-Shapur where he died in 769. In 785 al-Hadi, mindful of Jirjis' excellent services, invited his son Bukhtyishu' II, who had succeeded his father as head of the academy and hospital to go to Baghdad, but at court he had to face such determined opposition from Abu Quraysh the Queen's physician that for the sake of peace he was sent back to Jundi-Shapur. Under Harun ar-Rashid he was again summoned to court to treat the khalif for severe headaches, and later his son Jibra'il was brought to court and remained there until his death in 828–9. Whilst he was there the influence of the Barmakid wazir was making itself felt and efforts were being made to introduce to the Arabs the revived scientific learning derived from Greek sources, which was already spreading amongst the Syriac-speaking Christians. The Barmakid Yahya was an enthusiastic supporter of this revival of science with which he had been in touch in Marw, and was warmly supported by the Nestorian scholars of Jundi-Shapur.

Harun ar-Rashid became khalif in 786. He had been educated in Persia and under Persian influence at the hands of Yahya the Barmakid and throughout his reign showed strongly pro-Persian sympathies. He took great interest in science and literature, far beyond any of his predecessors, and the Hellenistic movement in Islam matured under his auspices. His reign was afterwards looked back upon as a golden age, but the khalifate had already begun to show signs of decay : in 800 he consented to the practical independence of the Aglabid governor of Qairawan in Libya, the beginning of a process of devolution which finally brought about the disintegration of the empire. Neither he nor any other of the 'Abbasid khalifs were able to extend their rule over Andalus, which had been a province under the 'Umayyads.

Influenced by his Barmakid minister Harun gave active support to the scholars who studied and translated Greek scientific works, sending out agents to purchase Greek manuscripts in the Roman Empire, a generous policy which brought a good deal of important material to Baghdad, and this was supplemented by similar generosity on the part of private persons who spent freely on manuscripts and translators. A good deal of the material thus obtained was medical and so appealed to the physicians of Jundi-Shapur, and this was rendered into Syriac as had been the case in former times, but before long Arabic versions made their appearance, at first translated from the Syriac, later directly from the Greek originals. The works of Aristotle were familiar in Syriac translations, and with them were commentaries and summaries, some composed in Syriac, others translated from the Greek. But at first the Aristotelian material was confined to the logical treatises. It was not until some time after the death of

Harun ar-Rashid that a serious and direct examination of Aristotelian philosophy was undertaken by Arab scholars. Derived through Syriac versions and commentaries the teaching of Aristotle was strongly tinctured with neo-Platonism, and that type of thought continued to colour Arabic philosophy to quite later times.

There seems reason to suppose that some of the earliest direct translations from the Greek was concerned with astronomy and mathematics. At an early date the *Sindhind*, an Indian treatise on astronomy and connected mathematics, based on Alexandrian teaching, was translated into Arabic, perhaps by means of a Persian version. It is said that the translators into Arabic were Ibrahim al-Fazari and Ya'qub ibn Tariq. Of the former of these Mas'udi says, " I will also cite the astronomer Ibrahim al-Fazari, author of the celebrated poem on the stars, astrology, and the study of the skies " (Mas'udi, *Muruj*, viii, 290), and then goes on to name him as one of al-Mansur's personal friends. The celebrated poem on the stars is not extant. He is said also to have been the first Arab to make an astrolabe. The son of this Ibrahim was Muhammad (d. between 769 and 806), who is sometimes mentioned as having been the translator. The date of a translation which is ascribed sometimes to the father, sometimes to the son, must be regarded as uncertain. Ya'qub ibn Tariq was a distinguished mathematician who is said to have been the author of a treatise on the sphere and another on the *karaja* or arc of 225, following the tradition of Archimedes who divided the circle into 96 degrees, and also to have drawn up astronomical tables. That the *Sindhind* was translated so early as al-Mansur is doubtful, but obviously the translation was well known to 'Abdallah Muhammad ibn Musa al-Khwarizmi, who made it the basis of his astronomical tables, but his work came some fifty years later, and the tables are now lost, but are cited and incorporated in later work by Maslama al-Majriti (*circ.* 1007). When tables are only known to us by being cited or incorporated in later work, we can never be sure how they have been touched up or improved, and how much remains of the original.

In order to understand and use the *Sindhind* it was found necessary to make translations of the *Almajest* (ἡ μεγίστη σύνταξις) of Ptolemy and Euclid's *Elements*, and these seem to have been translated directly from the Greek and to have been the earliest translation thus made. It is stated that it was made from a Syriac version, and this is not disproved by the absence of any such version surviving : Syriac literature is not rich in mathematical works. In favour of an early rendering from the Greek we have only the presumption that reference must have been made to the original to get an accurate rendering of the technical terms, a matter of the utmost importance in mathematical work. The Arabic versions were several times revised and corrected by comparison with the Greek text, so the earliest may have been made before Harun ar-Rashid, or in the early part of his reign. There is a tradition that the translations of Euclid and the Almajest were made at the suggestion of Ja'far the Barmakid, which would put them before 803, when the Barmakids fell into disgrace. If the observatory at Jundi-Shapur was in use before the time of an-Nahawandi (813–833), of which we cannot be certain, no doubt the necessary equipment in mathematics was available there and would be in Syriac. It is of course quite possible that the necessary mathematics were obtained from Indian works, not from Euclid or Ptolemy. The " Sons of Musa " had an observatory in Baghdad, but that would be after the time of Harun ar-Rashid.

Not much can be learned from the two astrologers who assisted al-Mansur in laying the foundations of Baghdad, though both of these are said to have produced mathematical, astronomical, or astrological works. One of these, an-Nawbakht (d. 776-7), is described as a convert from the Zoroastrian religion and a favourite of al-Mansur. He is said to have been the author of a work on judicial astrology and to have compiled astronomical tables, but of these works nothing survives. His son Abu Sahl al-Fadl an-Nawbakht (d. *circ.* 815) was Harun ar-Rashid's librarian and made translations from the Persian. The other astrologer, Mashallah, is said to have been a Jew of Marw whose name had originally been Misha, short for Manasseh (*Fihrist*, i, 273). Several of his works survive in Hebrew or Latin translations. Amongst these was a popular work on astronomy, not astrology.

It seems fairly certain that medical material came through a Syriac medium, direct translation from the Greek coming later. This may have been the case also with astronomical and mathematical material, but extant Syriac translations seem to be contemporaneous with the Arabic versions, not earlier, most indeed the work of Hunayn ibn Ishaq or his school.   It may be that mathematics and astronomy came through Indian authorities, not translations from the Greek but based upon Greek teaching, and translation from Greek into Syriac and Arabic came later when efforts were made to check and correct the available material.   Certainly the earliest Arab mathematicians, such as al-Khwarizmi, knew a great deal which does not appear in the Greek authors and much of which (but not all) can be traced to Indian workers. There are gaps in the chain of transmission which it is not easy to fill up.

CHAPTER XII

TRANSLATION INTO ARABIC

(1) THE FIRST TRANSLATORS

BAGHDAD was founded in 762. Harun ar-Rashid became khalif in 786 and in his reign Baghdad became the centre of a movement which aimed at translating Greek scientific material into Arabic. In the twenty-four years intervening between the foundation of the city and the accession of Harun ar-Rashid influences must have been at work to prompt this undertaking. Of such influences two were obvious, one radiating from Marw far away in Khurasan in the east, the other from Jundi-Shapur near at hand. Marw in Khurasan was indeed distant, but it had a great deal to do with early Baghdad. The 'Abbasids had been set upon the throne by a rebellion which had its source in Khurasan and which drew its chief support from that province. The Marw family of the Barmakids supplied the all-powerful ministers who guided and to a great extent controlled the 'Abbasid government. Many Persians, especially those of Khurasan, had flocked west to share in the triumph of the revolution and claim their share in its spoils. At the 'Abbasid court Persian influence very much thrust the Arab element into the background. The Persians were not modest about this : the Arabs had been arrogant, now the Persians repaid them with greater arrogance, deriding the Arabs as semi-barbarous nomads of the desert, parvenus without a history behind them, devoid of culture. This anti-Arab demonstration, open and plainly expressed, went by the name of the Shu'ubiya, an organized, virulent, and outspoken expression of anti-Arab feeling.

A typical figure of the times was Abu Muhammad ibn al-Muqaffa', a Persian who entered the service of 'Isa ibn 'Ali, uncle of the first two 'Abbasid khalifs, and became a convert to Islam, though many regarded his conversion as insincere. He translated from Pahlawi or Old Persian the book known as *Kalilag wa-Dimnag*, itself a translation of a Buddhist work brought from India by the Christian periodeutes Budh who had been sent to India to procure drugs, and with the drugs

brought back this book and the game of chess. Ibn al-Muqaffa'
produced a translation which is regarded as a model of classical
Arabic, and as such is still studied in schools. He also made
a translation of the Persian *Khudai-nama*, a biographical
history of the Persian kings, calling his Arabic version *Siyar
muluk al-'Ajam*. This work no longer exists, but it formed
the basis of Firdawsi's *Shah-nama* and many long extracts are
given in Ibn Qutaiba's '*Uyun al-akhbar*. In Arabic he composed
a treatise " on obedience due to kings " (*ad-durra al-yatima fi
ta'at a'-mulk*, printed Cairo, 1893 (?) and 1326, 1331 A.H.).
He also wrote several short treatises on " Adab ", etiquette,
duties of civil servants, and good manners, a favourite subject
in Old Persian literature. Living in Basra and feeling secure
in the protection of noble patrons he permitted himself many
impertinences towards Sufyan ibn Mu'awiya al-Muhallibi,
the city governor, jeering at him as " Ibn al-mughialina "
(son of the lascivious female), all of which Sufyan endured
in silence. After the rebellion of 'Abdallah against his nephew
al-Mansur the khalif agreed to pardon his uncle and Ibn
al-Muqaffa' was directed to draw up a formal letter of pardon
for the khalif to sign. In this letter he inserted " if at any time
the Commander of the Faithful act perfidiously towards his
uncle 'Abdallah ibn 'Ali, his wives shall be divorced from
him, his horses confiscated to the service of God, his slaves
set free, and Muslims absolved from allegiance to him ".
Al-Mansur read this draft and asked who had composed it.
On hearing that it was drawn up by Ibn al-Muqaffa' he said
nothing, but sent a letter to Sufyan telling him that he might
deal with the secretary as he saw fit. Various accounts are
given of the way in which the governor gratified his resentment
towards Ibn al-Muqaffa' by putting him to death, all of them
extremely cruel. This took place in 757-8 (*Ibn Khallikan*,
i, 432-3).

The cradle of the Shu'ubiya was Khurasan and its capital
Marw. Harun ar-Rashid himself was educated at Marw and
had strongly pro-Persian leanings. The astronomical records
kept under the Sasanid kings of Persia were continued under
the Arabs and were continued in Persian, not in Arabic, until
much later. From Marw came some of the earliest translators
of astronomical works, and it would seem that Khurasan was
the channel through which astronomical and mathematical

material came to Baghdad, for which very probably the Barmakid ministers, natives of Marw, were the agents. There was, it is true, an observatory at Jundi-Shapur, but we know little of its activity before the time of Ahmad an-Nahawandi (813–833), who made observations there some years after Harun's death. Some of the astronomical and mathematical material seems to have been obtained from India, derived from a Greek source in the first place, but probably it was transmitted to the Arabs through a Persian medium, though the actual Persian works whereby it was transmitted are no longer extant.

Jundi-Shapur was near Baghdad and under the 'Abbasid khalifs distinguished physicians were summoned thence to court. Successful in their professional work they remained in Baghdad as court physicians and became men of wealth and influence. Their success inspired other physicians to follow them and they, with scholars from Marw, formed a group under court patronage which became something very like an academy, a society of scholars rather than a teaching body. The men of Jundi-Shapur were accustomed to study Greek science in Syriac translations : gradually these Syriac versions were supplemented by Arabic ones, and finally the Arabic versions replaced them.

There is a legend that the *Sindhind*, the Hindu revised form of Brahmagupta's *Siddhanta*, was translated into Arabic as early as the reign of al-Mansur. It was an early translation, though probably not so early as that. But it proved useless as the Arabs could not understand it. It is related that Ja'far the Barmakid perceived the reason of this to be that the Arabs lacked the preliminary knowledge of geometry and astronomy necessary to follow it, and at his advice Harun ar-Rashid ordered a translation to be made of Euclid's *Elements* and Claudius Ptolemy's *megale* (*synaxis*). To this title the Arabs added the article al- and changed the *megale* into *megiste*, deliberately, it would appear, for Ya'qubi writing in 891 explained that "the meaning of al-majisti is ' the greatest book' " (Ya'qubim, ed. Houtsma, Leiden, 1883). Thus the work appears in Arabic as *Kitab al-Majisti*, which in medieval Latin became *magasiti*, presumably an attempted vocalizing of the unpointed *mjsty*. It does not appear that the translations of Euclid and Ptolemy were made until after the reign of Harun ar-Rashid, so the story that they were suggested by Ja'far ibn Barmak is dubious.

The translator of the *al-Majisti* is said to have been *al-Hajjaj ibn Yusuf ibn Matar al-Hasib*, who finished it about 827, which was well after the fall of the Barmakids and after the death of Harun ar-Rashid. The same translator is said to have made an Arabic version of Euclid's *Elements*, not including Book X which was later (about 910) translated with Pappus' commentary by Sa'id ad-Dimishqi. The translation of Euclid by al-Hajjaj with the commentary of an-Naziri (d. *circ.* 923), who also wrote a commentary on the al-Majisti, was published by T. O. Besthorn and J. L. Heiberg, *Euclidis elementa ex interpretatione al Hadschdschadschii cum commentario an Nazirii arab. et lat., ed. notisque* . . . Copenhagen, 1893. The earliest commentary on Euclid seems to have been that of al-'Abbas al-Jawhari (d. *circ.* 833). Another tradition represents the translation of the *al-Majisti* was made by *Sahl ibn Rabban at-Tabari*, a native of Marw and a Jew as his name *ibn Rabban* " the rabbi's son " denotes. Marw, one of the centres of Greek scholarship, had many Jewish neighbours who formed a colony of their own as was the Jewish custom, for they preferred to live in communities where the Jewish law could be observed. On the road betwen Marw and Balkh lay the city of Maymana which was at one time called al.yahudiya " the Jewish (city) ", but that name was changed to Maymana " the auspicious " at the request of its inhabitants who disliked the association with Jewry. This Sahl is described as having gone to Baghdad in the days of Harun ar-Rashid and having made the translation for him. He was a distinguished scholar and teacher of Marw who was known there as Barbun " the surpassing ". Some account of him is given by his son 'Ali ibn Sahl ibn Rabban at-Tabari (d. 850) in his great medical work *Firdaws al-Hikhma* " the Paradise of Wisdom " (ed. I. Siddiqi, Berlin, 1928). Yet another tradition represents the translation of *al-Majisti* as made by Sahl and revised by al-Hajjaj. This early version of the work was subsequently revised by Hunayn ibn Ishaq (below), later by Thabit ibn Qurra (below), then by Muhammad ibn Jabir ibn Sinan al-Battani (d. 929). Al-Hajjaj's translation of Euclid was revised by Qusta ibn Luqa about 912–13.

The earliest information which the Arabs obtained about

Aristotle from Syriac sources was confined to his logical works which had been translated and retranslated into Syriac, and on which several commentaries were accessible. The corpus of Aristotelian logic included the Categories, the Hermeneutics, the Prior Analytics, the Posterior Analytics, the Topics, the Sophistica, the Rhetoric, and the Politics, these last two works classed with the logical treatises by the Arabs. To these was added by *Yuhanna* (or *Yahya*) *ibn Batriq* about 815 another work, unfortunately a spurious one, the *Sirr al-asrar* or " secret of secrets ", which was accepted as Aristotelian. It is a work of miscellaneous contents, including physiognomy and dietetics.

Not long afterwards, about 835, a Christian of Emessa named *'Abd al-Masih ibn 'Aballah Wa'ima al-Himsi* translated another apocryphal work, the so-called " Theology of Aristotle ", really an abridged paraphrase of Plotinus, *Enneads*, iv–vi (cf. Fr. Dieterici, *Die sogennante Theologie des Aristoteles*, Leipzig, 1882).

About the same time lived *Abu Yahya al-Batriq* (d. between 798 and 806), who made an Arabic translation of an astrological work, the *Tetrabiblos* of Ptolemy. A commentary on this was written by 'Umar ibn al-Farrukhan (d. *circ.* 815), and a paraphrase by Muhammad ibn Jabir ibn Sinan al-Battani (d. 929).

Jibra'il I, the son of the otherwise unknown Bukhtyishu' I of Jundi-Shapur had attended al-Mansur, then retired to his own city and there finished his life. His son Bukhtyishu' II for a time acted as court physician to al-Hadi, but had to go back to Jundi-Shapur because of the opposition raised by the Queen's physician. He returned to the court of Baghdad under Harun ar-Rashid and attended both the khalif and his minister Ja'far the Barmakid. Before his death in 801 this Bukhtyishu' recommended his son Jibra'il II to the khalif, and he in due course became court physician. There is no evidence that the first two members of this family did anything to promote Greek science amongst the Arabs, but the second Jibra'il did, and as he acted in conjunction with Ja'far ibn Barmak it is obvious that he held an influential position in Baghdad even before his appointment as court physician. Bukhtyishu' died in 801 and then Jibra'il became the khalif's physician, after Harun's death in 808 continuing to serve his son al-Amin. But this led to his imprisonment when al-Ma'mun became master of Baghdad and all those who had been supporters of his brother al-Amin fell into disgrace. He was set free in 817 to attend the wazir Hasan ibn Sahl and lived without other disturbance until 829. He, no less than Ja'far ibn Barmak, was a patron and encourager of the work of translation from the Greek, a great admirer of Greek medical science, but was not himself responsible for any translation. He was the author of a *Kunnash* or medical compendium in Syriac in which he drew freely from Galen Hippocrates and Paul of Aegina ; this manual was long in use amongst Syriac-speaking practitioners and did a good deal to familiarize them with Greek medical teaching. The work is now lost, but some knowledge of it can be obtained from the tenth century Syriac lexicon of Bar Bahoul, who uses it to illustrate technical medical terms (Bar Bahoul, edited by R. Duval, Paris 1888–1898). It was largely at his suggestion that Harun ar-Rashid sent into the Roman Empire to obtain manuscripts and commissioned translations from the Greek. He and other contemporary patrons not only provided for Arabic translations but also encouraged the preparation of improved Syriac versions, for it is worth noting that a new and better series of translations into Syriac was being made at the same time that translation into Arabic was commenced. Translation into Syriac went on as long as the Jundi-Shapur academy was in existence.

The general conclusion is that the work of translation of scientific material began under Harun ar-Rashid with the encouragement of the wazir Ja'far ibn Barmak, and that this at first was especially concerned with mathematical and astronomical works, several of them translated by scholars from Ja'far's own city of Marw. The translation of medical works perhaps began a little later, and was associated with Jibra'il II. But there seem to have been some other translators not connected with the semi-official group gathered at court. Medical works came through Syriac versions in the first place and so did at least some of the astronomical and mathematical material, but in this latter direct reference to the Greek originals seems to have taken place earlier. This is as might be expected, for it was in mathematics that absolute accuracy in terminology was most important. Arabic lacked the technical terms used by Greek scientists. Sometimes the Greek terms were simply transliterated, but very often those terms show that they have passed through an Aramaic (Syriac) medium on their way, and this is more obvious in medical works than in mathematical and astronomical. As has been noted, the desire of more accurate scientific knowledge led to the preparation of more careful translations or the revision of existing versions, but it also resulted in the compilation of commentaries as well as original treatises based on the Greek authorities with citations illustrated and explained by original work. The encouragement of science became fashionable under Harun and many of the leading courtiers became patrons and spent freely on their scientific protégés. Not all of these may have been inspired by a pure love of science. When it became a fashion at court it is likely enough that many ambitious of advertising themselves found this a means of doing so. Outside court circles the scientific movement made small appeal. The Arabs generally took little interest in it : their learned men still spent their time in the study of Qur'an, jurisprudence, and grammar. So far, until the end of the reign of Harun ar-Rashid, no real work was done in the Aristotelian philosophy, Aristotle was treated only as an authority on logic.

Harun ar-Rashid died in 808, leaving the empire to his two sons al-Amin and al-Ma'mun, the former taking the western half with his capital at Baghdad, the other the eastern half with Marw as his capital. This naturally did not work and civil war between the two brothers followed inevitably. Al-Ma'mun's army, led by abler generals, obtained the upper hand, until in 812 under the leadership of Tahir it besieged Baghdad. This siege involved terrible sufferings and al-Amin was compelled to lay heavy requisitions on the citizens. At this the merchants entered into correspondence with Tahir. Discovering that he was betrayed al-Amin tried to escape and was on his way to make his submission to Tahir when he was found and murdered by some Persian free lances. These tragic events form the subject of an epic poem by al-Khuzaimi, a type of poem rare in Arabic.

At the death of al-Amin the whole empire fell into the hands of al-Ma'mun, but he preferred to remain at Marw and sent Hasan ibn Sahl to Baghdad as his deputy. Hasan's rule lasted six years, a period of tyranny and disorder gradually merging into anarchy, of which al-Ma'mun was kept in complete ignorance. At last the city revolted and elected Mansur ibn Mahdi governor until such time as al-Ma'mun could take over control in person. There was another reason why Baghdad was dissatisfied in addition to the tyrannical misrule of Hasan. Al-Ma'mun had invited the Shi'ite claimant to the throne, 'Ali ar-Rida, to Marw, received him with exceptional honour, and promised to make him his heir. This caused great offence at Baghdad which had no desire to be under Shi'ite rule.

At length the khalif was made aware of the critical state of affairs and warned that unless he went to Baghdad and took matters in hand for himself the khalifate would pass out of his hands. Thus warned he set out for Baghdad in 819, first disposing of 'Ali ar-Rida by poison. With him he took an extensive and extravagant court, as well as an army and also a select company of scientists, for he himself was deeply interested in scientific studies. At Baghdad he was welcomed with great rejoicings. He was a man of handsome presence, a thing which counts for much in oriental princes, generous, even lavish to extravagance in his expenditure and generally regarded as prudent, determined, of sound judgment, and clemency. According to the historians he was endowed with every grace and favour of an ideal prince. Educated in Marw in a neo-Hellenistic atmosphere, he applied philosophical principles to Muslim doctrines : no doubt others did the same, some of them men of exemplary piety, but they were careful to preserve external decorum by treating matters of religion with respect. Not so al-Ma'mun. He had a taste for discussing religious problems and this he did with considerable freedom, so that one of his courtiers once addressed him in jest as " Prince of Unbelievers ", a jest which was allowed to pass but its maker was never forgiven. Pro-Persian and anti-Arab, son of a Persian mother and married to a Persian wife, he had little in common with the narrow fanaticism of the typical Baghdadite. Unfortunately he was so far convinced of the rightness of the Mu'tazilite views that he determined to force them upon his subjects, selecting as a test point the question whether the Qur'an was, or was not, created. In 827 he published a decree penalizing any who did not agree that it was created and so not co-eternal with God. This decree was deeply resented as an innovation, for Islam has never

recognized the khalif as a religious teacher. The doctrines
of religion are defined, not by the State, but by those who
are learned in theology. As the penal was not successful,
al-Ma'mun reissued it in stricter terms with many peevish
complaints about the non-observance of his commands, and
established a *mihna* or inquisition before which any person
could be brought and examined as to his opinions, suffering
punishment if they differed from the officially authorized
rationalism. Under this law there were some martyrs and many
suffered imprisonment and other punishments, amongst them
Ahmad ibn Hanbal, a revered and greatly honoured tradi-
tionalist and jurist. All those who suffered were regarded as
saints.

Ten years after his arrival in Baghdad al-Ma'mun attempted
to repeat the experiment of the Greek geometer Eratosthenes
and measure the earth's arc. To do this he assembled a number
of scientists in the plain of Sinjar in Mesopotamia, west of
Mosul. The leading scientists thus gathered were Abu t-Taiyab
Sanad ibn 'Ali (d. after 860), who afterwards directed the
erection of the observatory in Baghdad, Yahya ibn Abi Mansur
al-Mai'muni, a freedman of al-Ma'mun's family, al-'Abbas
ibn Sa'id al-Jawhari (d. after 833), and 'Ali ibn 'Isa al-Astur-
labi. He divided these scientists into two parties which moved
apart until they saw a change of one degree in the elevation
of the pole. The distance travelled was then measured, and
it was found that one party had travelled 57 miles, the other
$58\frac{1}{2}$ miles, each mile reckoned as 4,000 " black cubits ", a
measure of length specially devised for this experiment. In
832 the experiment was repeated at Qasian, near Damascus.

When Jibra'il left Jundi-Shapur for Baghdad he was
succeeded as head of the academy and hospital there by
Abu Zakariah Yahya ibn Masawaih (d. 857), a Nestorian who
was the son of a druggist and had received his training as
a pupil of 'Isa b. Nun, who became Nestorian patriarch in
823. At that time medicine was in so great repute that it was
regarded as the foremost form of scientific education and
consequently it is common to find that Nestorian and Monophysite clergy in Asia often had a medical training rather
than one in litterae humaniores. But Ibn Masawaih left
Jundi-Shapur and went to Baghdad at Jibra'il's suggestion,
and was introduced at court as a skilful physician and one

learned in Greek medicine. He was the author of a treatise
on ophthalmology entitled *Daghal al-'ayn* " the disease of the
eye ", and also a collection of medical aphorisms *An-nawadir
aṭ-ṭibiyya*, which he dedicated to his pupil Hunayn ibn Ishaq.
This work attained great popularity and was translated into
Latin, but wrongly ascribed to St. John Damascene. In later
times Ibn Masawaih's treatise on the eye was so greatly
esteemed that it was selected as one of the set books for the
examination established by the Khalif al-Qahir (932–4) for
the licence to practise medicine, an examination at first under
the direction of Sinan ibn Thabit. There is also an " Instruction for the examination of oculists " which is ascribed to him,
but it is simply a cram book based on the *Daghal al-'ayn*
probably a later compilation made for the use of examination
candidates.   The *Daghal al-'ayn* " is the earliest treatise on
ophthalmology, the Greek, Syriac, and other special text-
books being lost.   It is written in bad Arabic, with many
Greek, Syriac, and Persian technical terms, a rather confusing
compilation without system, and doubtless intermixed with
later interpolations. One complete MS. is extant in Taimur
Pasha's library (Cairo), another in Leningrad " (M. Meyerhof,
The Book of the Ten Treatises, Cairo, 1928, ix–x). Analysis and
extracts of this work in German by M. Meyerhof and C.
Preufer, *Die Augenheilkunde des Juhanna ibn Masawaih*, in *Der
Islam*, vi, 1915, pp. 217–256.

(2) HUNAYN IBN ISHAQ

The most celebrated of all translators of Greek scientific
works into Arabic was *Hunayn ibn Ishaq al-'Abadi* (d. 873 or 877).
The outline of his life and work are well known from his autobiography written in the form of letters to 'Ali ibn Yahya in
875. (Text from two manuscripts in the Aya Sofia Mosque
at Stambul, ed. with translation by G. Bergeström, Leipzig,
1925.)   He was a native of Hira, the son of a Christian
(Nestorian) druggist.   In later life he learned Arabic, so
presumably he did not belong to the ruling class of Hira which
was Arabic-speaking, and this is endorsed by his name 'Abadi,
which shows that he belonged to the subject people of Hira.
As a young man he attended the lectures of Ibn Masawaih
(above) at Jundi-Shapur, and so far earned the approval of

his teacher that he was made his dispenser. But later he annoyed Ibn Masawaih by asking too many questions in class, and at least his teacher lost patience and said : " What have the people of Hira to do with medicine ?—go and change money in the streets," and drove him out weeping (Ibn al-Qifti, 174). Expelled from the academy Hunayn went away to " the land of the Greeks " and there obtained a sound knowledge of the Greek language and familiarity with textual criticism such as had been developed in Alexandria. In due course he returned and settled for a time at Basra where he studied Arabic under Khalid ibn Ahmad then, some time before 826, proceeded to Baghdad where he obtained the patronage of Jibra'il and for him prepared translations of some of Galen's works. Harun ar-Rashid died in 808 and al-Ma'mun succeeded in 813, after the brief and stormy reign of al-Amin, so that Hunayn's activities belong to a period later than Harun ar-Rashid. The excellence of his translations, far surpassing any previous work of the sort, greatly impressed Jibra'il who then introduced him to the three " Sons of Musa ", wealthy patrons of learning. Their father, Musa ibn Shakir, after a life spent in the lucrative profession of a brigand in Khurasan, had reformed and been pardoned, then settled down to spend his declining years in cultured leisure. He entrusted his sons to the Khalif al-Ma'mun, who appointed Ishaq ibn Ibrahim, and later Yahya ibn Abi Mansur to be their teachers, and from those preceptors they received a training in mathematics. They were not so much interested in medicine, but patronized Hunayn chiefly because of his excellence as a translator. Of these " Sons of Musa " the eldest Muhammad rose to high office under the Khalif al-Motadid (892–932), and distinguished himself in astronomy and geometry, a second son Ahmad excelled in mechanics, and the third son Hasan attained celebrity in geometry. They had a house in Baghdad near the Bab at-Taq, the gate at the eastern end of the main bridge over the Tigris, opening into the great market street of East Baghdad, and there they built an observatory where they made observations during the years 850–870. To them we owe a treatise on plane and spherical geometry, a collection of geometrical problems and a manual of geometry which was translated into Latin by Gerhard of Cremona (d. 1187) as " Liber Trium Fratrum de geometria " (ed. M. Curtze in *Nova Acta d. Kais. Leop. Carol. deustscen Akad. Naturforscher*, xlix, 109–167), which long held its own as an introduction to geometry. They were generous patrons of scientific research and according to Ibn Abi Usaibi'a spent at one time an average of 500 dinars (say £200) a month on their scientific protégés.

The " Sons of Musa " introduced Hunayn to the Khalif al-Ma'mun some time before Jibra'il's death in 828–9, and apparently at Jibra'il's suggestion the khalif founded an academy which he called the " House of Wisdom " (*Dar al-hikhma*) as an institution where the preparation of translations from Greek scientists would be made and circulated amongst the Arabs, placing Hunayn in charge. From that time forwards the work of translation went on steadily, and before long Arab students found themselves equipped with the greater part of the works of Galen, Hippocrates, Ptolemy, Euclid, Aristotle, and various other Greek authorities. The work of translation was twofold, versions were made in Arabic and also in Syriac, these latter to replace the defective translations already in use. Ibn Masawaih, the teacher who had expelled Hunayn from Jundi-Shapur, was reconciled to him and became his warm supporter. Hunayn had many other friends and clients, mostly physicians of Jundi-Shapur and those who had removed to Baghdad and used the Arabic language, like Salmawaih ibn Bunan an alumnus of Jundi-Shapur who became court physician to al-Mu'tasim in 832. All these were better translations than had been known in the past and were made from good Greek manuscripts, many of them procured by agents of the khalif who were sent into the Roman Empire and empowered to spend considerable sums on the purchase of the best codices.

Altogether Hunayn translated into Syriac twenty books of Galen, two for Bukhtyishu' Jibra'il's son, two for Salmawaih ibn Bunan, one for Jibra'il, and one for Ibn Masawaih, and also revised the sixteen translations made by Sergius of Rashayn. He translated fourteen treatises into Arabic, three for Muhammad, one for Ahmad, sons of Musa. He and his assistants produced versions both in Syriac and Arabic, though no doubt some of his staff excelled in one language rather than the other. Most of the translators of the next generation received their training from Hunayn or his pupils, so that he stands out as the leading translator of the better type, though some of his versions were afterwards revised by later writers.

The complete curriculum of the medical school of Alexandria was thus made available for Arab students. This included a select series of the treatises of Galen which was :—

1. De sectis.
2. Ars medica.
3. De pulsibus ad tirones.
4. Ad Glauconem de medendi methodo.
5. De ossibus ad tirones.
6. De musculorum dissectione.
7. De nervorum dissectione.
8. De venarum arteriumque dissectione.
9. De elementis secundum Hippocratem.
10. De temperamentis.
11. De facultatibus naturalibus.
12. De causis et symptomatibus.
13. De locis affectis.
14. De pulsibus (four treatises).
15. De typis (febrium).
16. De crisibus.
17. De diebus decretoriis.
18. Methodus medendi.

The range and method of Hunayn's work is known to us from his autobiography, the *Risalat Hunayn ibn Ishaq*, letters written to 'Ali ibn Yahya in 865, of which the text with translation has been published from two manuscripts in the Aya Sophia Mosque at Stamboul, by G. Bergesträsser, Leipzig, 1925, a work which has been analysed by Dr. Meyerhof in *Isis*, viii (1926), 685–724.

Al-Ma'mun's reign came to an end in 833 and he was succeeded by his son al-Mu'tasim (833–842), who found it difficult to control the populace of Baghdad and formed a guard of Turkish slave-soldiers. But this body-guard, holding a privileged position, soon became insubordinate and many complaints were made about their conduct. At last al-Mu'tasim in 836 removed himself and his court to Samarra, and there the khalifs reigned until 892. These disorders affected scholarship adversely and the " House of Wisdom " fell into decay which was not checked during the brief reign of Wathiq (842–7).

As Wathiq's son was too young to occupy the throne his brother Mutawakkil (847–861) was invested with the khalifate. His accession made a great change. The previous khalifs had been tolerant in religion, al-Ma'mun was generally regarded as a free-thinker. But Mutawakkil was of the strictest orthodoxy and fanatical in his orthodoxy, possibly afraid of the disaffected attitude of the Syrian Christians. He was of sadistic temperament, mischievous and capriciously cruel. Though not himself a scholar like al-Ma'mun, he was a patron of science and scholarship and reopened the Dar al-Hikhma, granting it fresh endowments. The best work of translation was done during his reign, as the training of the staff and experience were bearing fruit.

Mutawakkil's personal relations with Hunayn were chequered. It is related that the khalif told him to prepare poison for his enemies and, on Hunayn's refusal to do so, cast him into prison. Not long afterwards he was released and Mutawakkil explained that he had only desired to test his loyalty to the traditional standards of medical practice. Then a Nestorian physician named Isra'il ibn Zakariya at-Taifuri, or else his friend Bukhtyishu', denounced him as a heretic, that is a heretic from the Nestorian standard, for Hunayn had never conformed to Islam. The Nestorian Church, like other tolerated religious communities, was self-governing in its private affairs and could punish heretics and other offenders, though the khalif quite gratuitously comes into the story. It is said that Mutawakkil ordered Hunayn to spit on a picture of the Holy Theotokos and on his refusal handed him over to the Nestorian Catholicos Theodosius who imprisoned and scourged him. The implication seems to be that the khalif invited him to repudiate Christianity, and when he refused to do so handed him over to the Nestorian Catholicos for punishment. Just possibly this vague and confused story contains an echo of the Iconoclastic controversy which at that time was disturbing the Eastern Church. Mutawakkil further confiscated Hunayn's property, including his library, a loss which he felt sorely. After four months he was set free because of a remarkable cure following his treatment of a court dignitary, and his goods and library were restored. The whole matter sounds very much like an intrigue amongst the court physicians, as on his release the other court physicians had to pay him 10,000 dirhams compensation.

After his release he lived another twenty years, which he employed in making translations and correcting those made by others. In 861 Mutawakkil was murdered by his Turkish guards at his son's instigation. Hunayn enjoyed the favour of that son Montasir (861-2), and of his successors Mosta'in (862-6), Mo'tazz (866-9), Muhtadi (869-870), and Mu'tamid (870-892), and was engaged in making a translation of Galen's *De constitutione artis medicae* at the time of his death, which took place in 873 according to the *Fihrist*, or in 877 according to Ibn Abi Usaibi'a, who is often inaccurate in his chronology. According to I.A.U., Hunayn was the author of more than a hundred original works, but only a few of these are extant. Hunayn, the greatest of the translators, must be reckoned to the credit of Jundi-Shapur, although his fuller and more accurate knowledge was gained by his studies in the " land of the Greeks ", for those travels and studies were prompted and directed by what he had learned at Jundi-Shapur under Ibn Masawaih.

Although Mutawakkil was bigoted, fanatical, and sadistic, he was a generous patron of scientific research and is generally reckoned as having re-endowed the " House of Wisdom ", which probably means that it was reopened after the disturbed period which followed al-Ma'mun's death and its endowments restored to it. The best work of this academy was done under Mutawakkil, for by that time experience told and Hunayn was surrounded by well-trained pupils.

Amongst those who worked with Hunayn must be noted his son *Ishaq*, who died in November, 910 or 911, and his nephew *Hubaysh ibn al-Hasan*, who was at work in the days of Mutawakkil. He translated Greek texts of Hippocrates and the botanical work of Dioscorides which became the basis of the Arab pharmacopœia (*infra*). It is noteworthy that most of the names of planys in Arabic show that they have passed through an Aramaic (Syriac) medium (cf. Loew, *Aramäische Pflanzennamen*, 1881).

Another noteworthy pupil was '*Isa ibn Yahya ibn Ibrahim* who was a translator of Greek medical works into Arabic.

M

Almost all the leading scientists of the succeeding generation were pupils of Hunayn.

Although Hubaysh is given as the translator of Dioscorides, the current Arabic version is more commonly ascribed to Hunayn's pupil *Staphanos ibn Basilos*, who translated the work into Syriac, and this Syriac version was then translated into Arabic by Hunayn himself (or Hubaysh) for Muhammad, one of the " Sons of Musa ". But another independent version of Dioscorides was afterwards made in Spain (cf. below).

(3) OTHER TRANSLATORS

About 908 the Christian priest *Yusuf al-Khuri al-Qass* translated Archimedes' (lost) work on triangles from a Syriac version, and this was afterwards revised by Thabit ibn Qurra. He also made an Arabic translation of Galen's *De simplicibus temperamentis et facultatibus*, which was afterwards revised by Hunayn ibn Ishaq.

About the same time lived *Qusta ibn Luqa al-Ba'lbakki* (3. 912–13), a Syrian Christian who translated Hypsicles, afterwards revised by al-Kindi, Theodosius' *Sphaerica*, which was afterwards revised by Thabit ibn Qurra, Heron's Mechanics, Autolycus, Theophrastus' *Meteora*, Galen's catalogue of his books, John Philoponus on the Physics of Aristotle and several other works, and also revised the existing translation of Euclid.

Abu Bishr Matta ibn Yunus al-Qanna'i (d. 940) was responsible for a translation of the *Poetica* of Aristotle.

Medical and logical works were translated also by the Monophysite *Abu Zakariya Yahya ibn 'Adi al-Mantiqi* " the logician " (d. 974), amongst them the *Prolegomena* of Ammonius, an introduction to Porphyry's *Isagoge*.

To these may be added the late translator *Al-Hunayn ibn Ibrahim ibn al-Hasan ibn Khurshid at-Tabari an-Natili* (d. 990) also the Monophysite *Abu 'Ali 'Isa ibn Ishaq ibn Zer'a* (d. 16th April, 1008), who prepared versions of medical and philosophical works. With these the series of translators in Asia comes to an end. After this the work changes to commentary and exposition, occasionaly revising earlier translations.

A final phase of translation appears in Andalus, the Muslim occupied Spain. There the fugitive 'Umayyad prince

'Abdarrahman had established an independent kingdom in 755. The eighth prince of that Andalusian state 'Abdarrahman III in 929 adopted the title Khalif and so from 929 to 978 there were khalifs of Cordova, usually with strained relations with the 'Abbasids in the east, but friendly with the Emperor of Byzantium who was their enemy. In 949 the Byzantine Emperor Constantine VII sent an embassy to Cordova and amongst the presents he sent to 'Abdarrahman was a copy of Dioscorides in Greek with painted pictures of the many plants described in the text. This book attracted much attention, but no one in Cordova could read Greek, so the Khalif in thanking the Emperor begged him to send someone who could translate and explain the work. In 951 the Emperor sent a monk named Nicolas, who was able to speak Arabic, and he not only made translations of Dioscorides and other Greek works, but began teaching the Greek language, his lectures arousing great enthusiasm and being attended by many court officials, including Hasdai ibn Shaprut, the Jewish wazir. Translations of Dioscorides already existed, that of Hunayn ibn Ishaq from the Syriac version of his pupil Stephenos ibn Basilos, and the version made by an-Natali for the Prince Abu 'Ali as-Sanjuri. But Nicolas made an improved translation in which pains were taken to identify the plants described, thus laying the foundation of a serious study of botany which very quickly bore fruit in the work of Abu Dawud Sulaiman ibn Juljul (*circ.* 1000), physician to 'Abdarrahman's successor Hisham II, who wrote a supplement to Dioscorides describing a number of plants found in Spain, a land peculiarly rich and varied in its flora, but not known to the Greek author. Although there was a very productive cultural harvest in Andalus and the reign of 'Abdarrahman III was the golden age of Andalusian culture, there does not appear to have been any further output of translations from the Greek there. The Andalusian version of Dioscorides as made by Nicolas exists in a Bodleian manuscript. Apparently the older version prepared by Hunayn ibn Ishaq or an-Natali was quite unknown in Spain.

(4) THABIT IBN QURRA

Thabit ibn Qurra is prominent amongst those who revised and corrected Arabic translations of mathematical and astronomical works, and introduces a new source of pro-Greek interest. He was a native of the town of Harran, the ancient Charrae, where men adhered steadfastly to their ancient paganism, although the deities worshipped there bore names borrowed from the Greek pantheon. It was in the midst of Syriac Christian culture, between Edessa and Rashayn, situated on the Belias, a minor tributary of the Upper Euphrates. It was famous for the purity of the Aramaic spoken there, and this was sometimes attributed to its comparative freedom from Jewish or Christian influences, though in fact there was a Christian bishop who claimed Harran as his see and presumably there was a Christian congregation there. It seems to have been in touch with the renaissance of Greek learning which affected both the Nestorian and Monophysite churches and its thought was strongly tinctured with neo-Platonism.

Our knowledge of the ancient religion of Harran is chiefly gleaned from the observations of ad-Dimishqi, who died in A.D. 1327, long after the city had passed into obscurity and who could only have had traditional information about its religion. His information is summarized in Chwolson's *Die Ssabier und der Ssabismus*, ii, 280–411. From that we learn that the Harranians had five great temples dedicated respectively to the First Cause, the First Reason, the Ruler of the World, Form, and Soul. There were seven other temples dedicated to the seven planets. It was an anomaly for a pagan city to enjoy religious freedom under Muslim rule and non-interference was not due to the city being obscure as it was the capital of the province of Diyar Mudar and under the last 'Umayyad khalif Marwan II it was the residence of the court and government administration. The *Fihrist* relates a story that al-Ma'mun towards the end of his reign passed by Harran on a military expedition, and he and his officers were astonished at the strange and uncouth appearance of the townsmen. He asked who they were, and was shocked to learn that they were pagans. This implies that Harran was unknown to Muslims generally and a remote isolated district, which is not true. Al-Ma'mun ordered the people to adopt one of the recognized religions, Islam, Judaism, Christianity, or Mazdeanism, before he came back that way. He never did come back, but the people were alarmed at his threats and many of them conformed to Islam or Christianity : Mazdeanism seems to have ceased to make converts by then ; but others adhered to their paganism and sought a way to escape the Khalif's anger. A certain lawyer offered to show them a possible way of doing so for a consideration, and when they had paid him his fee he advised them to claim to be Sabaeans (Ṣabi'a), as those are mentioned in the Qur'an as one of the " peoples of the Book " (Qur., 2, 59 ; 22, 17 ; 5, 73), and no one knew who the Sabaeans were. The story is obviously apocryphal : the Harranites could not have been so little known in the days of al-Ma'mun as his father Harun ar-Rashid had already put pressure on them as heretics and their city had been the seat of government · under Marwan II. The story is an attempt to explain how the Harranites came to be called Sabaeans, a name which we now recognize as not belonging to them. The real Sabaeans were a people of South Arabia, with whom Harran had no concern. But the Mandaeans of the Lower Euphrates, the Haemerobaptists of the Christian fathers and the rabbinical writers who earned the title of " baptists " from their frequent and punctilious ablutions, were in Aramaic called Ṣaba'in from the root ṢB' " ummerse ". Those Mandaeans were Gnostics who inclined to astrological beliefs, possibly actual star-worshippers. The people of Harran were not Gnostics, but they had temples dedicated to the planets, which gave some colour to the confusion between them and the Mandaeans. Harranite neo-Platonism might possibly be confused with Gnostic beliefs. It is characteristic that the Harranites claimed that their religion had come to them from Hermes. It is an interesting instance, though not a unique one, of the way in which the Muslim law was sometimes evaded.

Thabit ibn Qurra (d. 901) " was originally a money-changer in the market of Harran, and when he turned to philosophy he made wonderful progress and became expert in three languages, Greek, Syriac, and Arabic. . . . In Arabic he composed about 150 works on logic, mathematics, astronomy, and medicine, and in Syriac he wrote another fifteen books " (Bar Hebraeus, *Chron.*, x, 176). About 872 he was excommunicated by the High Priest of Harran, unfortunately we know nothing about the ecclesiastical discipline of Harran— and sent to Kafartutha, near Dara, but he remained staunch to his religion. " Our fathers," he said, " by the help of God stood firm and spoke boldly, so this favoured town never was polluted by the error of Nazareth (Christianity), and we are their heirs and transmitters of paganism in these days : fortunate is he who bears his burden in hope strengthened by paganism (ibid.). He maintained that it was the pagans who first cultivated the land, founded cities, made ports, and discovered science (ibid.)." After wanderings in various lands he met Muhammad, one of the " Sons of Musa ", who recognized his scholarship and took him to Baghdad where he did most of his work. He made translations of Apollonius, Archimedes, Euclid, Ptolemy, and Theodosius, or revised existing translations. He also composed several works on astronomy and mathematics. It has been supposed that he was responsible for the extremely mechanical form in which Ptolemy's cosmography was presented to the Arabs, but that hardly seems justified. In mathematics he introduced the theory of "amicable numbers", a Chinese idea. Such numbers are those in which one is the sum of the factors in the other. Thus if $p = 3 \cdot 2^n - 1$, $q = 3 \cdot 2^{n-1} - 1$, and $r = 9 \cdot 2^{2n-1} - 1$, assuming that n is a whole number, then $a = 2^n pq$, and $b = 2^n r$ are amicable numbers. Suppose $n = 2$, then $p = 3 \cdot 2^n - 1 - 11$: $q = 3 \cdot 2^{n-1} - 1 = 5$: $r = 9 \cdot 2^{2n-1} - 1 = 71$: so the amicable numbers are $a = 320$, $b = 284$. Nothing very much results from this investigation, but it was continued by Maslama al-Majriti and a few other Arab mathematicians.

Thabit had a son Abu Sa'id who became physician to the Khalif al-Qahir. He also was a pagan, but the khalif tried to convert him to Islam and fell into the habit of using the most bloodthirsty threats to force him to do whatever he wanted, until the unhappy physician fled to Khurasan and remained there until al-Qahir was dead. Then he returned to Baghdad and lived there until his own death in 943. Thabit had many pupils, one of whom a Christian named 'Isa ibn Asd translated into Arabic various works which Thabit had composed in Syriac.

About 932–4 the city of Harran was destroyed either by the 'Alids, as Hamawi says, or by Egyptian invaders as Dimishqi asserts. The contemporary historian, John of Antioch, describes this destruction.

In 975 Abu Ishaq ibn Hilal, secretary to the khalifs Muti'
and Ta'i', obtained a decree granting religious toleration to
the Sabaeans of Harran, of whom there were many in Baghdad—
some were still there in the eleventh century—one of whom
the most distinguished were the mathematician Abu Ja'far
al-Khazin, a convert to Islam, and Ibn al-Wahshiya, author
of a work known as " the Nabataean Agriculture " (Kitab
al-falaha an-nabatiya), which pretended to be a translation
from ancient Babylonian. This work was finished in 904 :
it is a collection of popular beliefs, superstitions, and legends.
It gives no real botanical information but simply aims at
proving that the ancient Babylonian civilization existed ages
before the rise of the Arabs whose culture was a comparatively
recent and inferior one. In fact it is an example of the strong
anti-Arab animus characteristic of the early 'Abbasid period.
The work had no influence on the development of intellectual
culture amongst the Muslim Arabs.

After its destruction in 932–4 Harran was rebuilt, but
destroyed again in 1032 when only the great Temple of the
Moon was left standing. After these misfortunes it still lingered
on and was visited by Ibn Jubayr in 1184, but in 1332 Abu
l-Feda found only a decaying village on its site.

CHAPTER XIII

THE ARAB PHILOSOPHERS

ARISTOTLE dominated the later school of Alexandria and his influence inevitably passed over to the Christian world and so to Islam. The Syriac study of Aristotle took form in the school of Edessa in the fifth century, his teaching then being chiefly confined to logic. With Aristotle's logical works were associated the *Isagoge* of Porphyry and his philosophy generally by the summary of the Syrian writer Damascius. Fuller study was reached by the use of commentaries, first by that of the Syriac Probus, then by the Alexandrians Ammonius and John Philoponus. Now it will be noted that these works used to interpret Aristotle were predominantly neo-Platonic, and that neo-Platonic strain remained in Arabic philosophy and influenced both it and Muslim theology. This influence was further increased by the acceptance of the abridgment of Plotinus' *Enneads*, iv–vi, as "the Theology of Aristotle" and so a genuinely Aristotelian work.

The fame of Aristotle spread amongst the Muslims as soon as they began to turn their attention to Greek scientific material, but for some time his actual teaching, very imperfectly reproduced at second hand, was all that was accessible to them. When they knew it better they found it not altogether to their liking, especially in the doctrine of the eternity of the universe which contradicted the Qur'anic teaching of creation, the denial of a special providence which conflicted with the idea of a divine control of affairs as taught in the Qur'an, and the denial of the resurrection of the body, all of which seemed to the orthodox little better than blasphemy. At first Aristotle was accepted only as a logician, but afterwards translations were made of some of the treatises on natural science, a very unsatisfactory one of the Metaphysics, and to these were added several spurious works, though of these the only definitely tendencious one was the so-called Theology.

Aristotelian study proper began with *Abu Yusuf Ya'qub ibn Ishaq al-Kindi* (d. after 873), commonly known as "the philosopher of the Arabs", was of pure Arab birth, though the

Chahar Maqala strangely refers to him as a Jew, in spite of the emphasis always laid on the purity of his Arab descent. He was born at Kufa where his father was governor, educated at Basra and Baghdad, and was still alive in 873. At first he worked as a translator and did not undertake any original work until he had proved his competence in making translations of Greek philosophical and scientific works. He became entirely devoted to the teaching of Aristotle and is generally regarded as the first of the line of Arab philosophers who professedly followed the neo-Aristotelian school. It was to such that the Muslims applied the name of " philosophers ", using the term to designate those whom they regarded as members of a sect definitely unorthodox in its tendencies. Al-Kindi's own speculations in theology were of the Mu'tazilite or rationalist type prevalent at al-Ma'mun's court and which that prince tried to enforce generally by issuing a decree asserting the Qur'an to be created, not co-eternal with God. Al-Ma'mun made him tutor to the prince who in due course ascended the throne as al-Mu'tasim (833–847), and it is said that for him al-Kindi translated the so-called " Theology of Aristotle ", although that translation was also attributed to 'Abd al-Masih al-Himsi and with greater probability, for al-Himsi was a Syrian Christian and it was in Syria that the work received its readiest welcome. Possibly it was translated by al-Himsi and revised by al-Kindi. Certainly al-Kindi accepted it as a genuine Aristotelian work and adopted its teaching, which shows a type of mystical theology easily inclining towards pantheism, indeed pantheistic tendencies constantly showed themselves in Arabic Aristotelianism. Like other rationalists al-Kindi fell under suspicion at the accession of the rigidly orthodox Mutawakkil in 847 and was disciplined by the confiscation of his library like Hunayn ibn Ishaq, but after a while it was restored to him.

His chief importance lay in his definite acceptance of Aristotle as " the Philosopher ", no longer simply as a teacher of logic. He professed to be his follower and took him as authoritative, practically inspired, teacher, and in this was the founder of the Arab Aristotelian school, though his actual work lay chiefly in translating and introducing to the Arabs the teaching of the Philosopher instead of the vague and inaccurate notions they had gathered and exaggerated in the process from Syriac exponents. In the Arabic Aristotelian school the teaching of Aristotle was accepted even when in conflict with the literal statements of the Qur'an. It was regarded as truth which was only intelligible to the enlightened, whilst the Qur'an and orthodox doctrine generally served well enough for the unlettered and was best adapted for them. Some followers of this school went further and held that the Qur'an had an esoteric meaning disclosed only to the discerning, and that that esoteric meaning agreed with the teaching of Aristotle. It was the familiar problem, granted that science and revelation are both true, they must somehow agree together although they seem to contradict one another.

It was, however, *Abu Nasr Muhammad al-Farabi* (d. 950) at the court of Sayf ad-Dawla, at Aleppo, who really shaped the philosophical teaching of Arabic Aristotelianism, basing his work on the better knowledge of the text of Aristotle made accessible by the labours of al-Kindi. Al-Farabi was of a Turkish family of Transoxiana, but had studied in Baghdad under the Christian physician Yuhanna ibn Hailam and Abu Bishr Matta, already mentioned as a translator. He was a commentator on Aristotle and built up a system of philosophy from Aristotelian and neo-Platonic material, this latter then generally accepted as the correct interpretation of " the Philosopher's " teaching, which resulted in a kind of Muslim neo-Platonism. From this he came to be known as " the second teacher ", that is to say, the authority next after Aristotle. He accepted the Qur'an as true, but maintained that philosophy also was true, so the two must agree : in so far as they appear not to agree steps must be taken to reconcile them, for truth must be consistent and apparent inconsistencies can be explained away.

He assumed that Plato and Aristotle were at one. This was then the accepted view, and as Plato was known in the neo-Platonic form as interpreted by Porphyry, the resultant system was very strongly tinctured with neo-Platonism. " The more pious added the third element of the Qur'an, and it must remain a marvel and a magnificent testimonial to their skill and patience that they even got so far as they did, and that the whole movement did not end in simple lunacy. That al-Farabi should have been so incisive a writer, so wide a thinker and student : that Ibn Sina should have been so keen and clear a scientist and logician, that Ibn Rushd should have known—really known—and commented his Aristotle as he did, shows that the human brain, after all, is a sane brain and has the power of unconsciously rejecting and throwing out nonsense and falsehood " (D. B. Macdonald, *Development of Muslim Theology*, 163). It is significant that almost all the great scientists and philosophers of the Arabs were classed as Aristotelians tracing their intellectual descent from al-Kindi and al-Farabi and most of them professed to belong to that school.

But al-Kindi's more accurate study of Aristotle had not entirely disposed of the older inaccurate pseudo-Aristotelianism which had prevailed amongst the imperfectly informed Arabs of an earlier day. Probably in the opening years of the tenth century and in Baghdad there was gathered a group of men who called themselves the *Ikhwan as-Safa* " the Brotherhood of Purity " or " the Sincere Brethren ", but is more probably intended to express the term " philosophers ", at a time when the recent accession to power of the Buwayhid dictators produced a temporary experience of toleration and free thought. Somewhere about A.D. 980 this group produced a body of epistles or essays which aimed at being a complete encyclopædia of philosophy and science. These essays are 52 in number : the first fourteen deal with mathematics and logic, 15–31 with natural science, 32–41 with metaphysics, the remainder with mystic theology, astrology, and magic. Epistle 45 describes the organization and guiding principles of the brotherhood. Very commonly the Imam Ahmad is given as the author of this work, but Shahruzi names five contributors, Abu Hasan 'Ali b. Harun az-Zinjani, Abu Ahmad an-Nahajuri (or Mihrajani), Abu Sulaiman Muhammad ibn Nasr al-Busti (or al-Muqaddisi), al-'Awfi, and Zayd ibn Rifa'a. These letters were produced in or near Basra or Baghdad. The contents show a kind of obscure and crude type of Aristotelianism, such as was current in the earlier period of the revival of Greek science, before al-Kindi had set a more accurate standard, but references are made to older philosophies, to Hermes, Pythagoras, Socrates, and Plato, all confused and vague. Aristotle appears chiefly as a logician : the " Theology of Aristotle " and the " Book of the Apple " are accepted as genuine Aristotelian works. No reference is made to al-Kindi or his work, but Abu Ma'shar and other eighth or ninth century writers are quoted. There is no trace of the influence of al-Kindi. The doctrine contained in these letters is eclectic, the world is described as an emanation from God, the human soul as of celestial origin and striving to return to God and to be absorbed in Him, a consummation to be attained by wisdom, the *Gnosis* of Gnostic and neo-Platonic writers. The Qur'anis interpreted allegorically, and reference is made to the Christian and Jewish scriptures, which are treated in a similar way. This teaching shows distinctly Shi'ite, probably Isma'ilian, tendencies, but the language in which it is expressed is involved and obscure, perhaps intentionally so with the intention of veiling spiritual teaching from the profane. The Batini or allegorical movement had its roots in older non-Muslim thought, and presumably had survived in Lower Mesopotamia where were many ancient creeds, all more or less mixed up with politically subversive movements : this was the area in which the Khalif al-Mahdi had tried to suppress the Zindiqs or " atheists ", and in which the Qarmates afterwards had their beginnings, the home of the Isma'ilians, in any case definitely anti-'Abbasid and anti-Arab. In Islam this kind of Batini thought was strongest in the Isma'ilian sect, it had strong Gnostic tendencies and laid great stress on the spiritual and esoteric, as against the exoteric (Lewis, *Origins of Isma'ilism*, Camb., 1940, 44 sq.). This type of thought is interesting as it represents the " wisdom " cherished by the Isma'ilians, by their adherents in the Fatimid khalifate in Egypt and later by the Assassins of Central Asia and Syria, offshoots of the Fatimids, and presumably by the Druzes of the Lebanon. Though very far removed from the natural line of Islamic thought it still forms a living and vigorous branch of Islam, though it is not Arab.

Reference has already been made to the attitude which was adopted by the " philosophers " towards the Qur'an and orthodox doctrine generally. This is best illustrated by reference to the philosophical romance of *Haiy ibn Yuqsan* " The Living One son of the Wakeful ", composed by the Andalusian philosopher Abu Bakr Muhammad ibn Tufayl, who died in Maghrab (Morocco) in 1185-8. This book pictures two islands, one densely peopled, the other believed to be uninhabited. On the former are ordinary people living conventional lives and satisfied with the customary observances of the precepts of religion. Amongst them are two prominent characters, Asal and Salaman, who by self-discipline have raised themselves to a higher plane. Salaman outwardly adapts himself to conventional religion, but Asal tries to discover deeper spiritual truths by meditation : to do this the better he removes to the other island where he finds one occupant Haiy ibn Yuqsan who has lived there in solitude from infancy and by the innate powers of his mind has developed a lofty philosophy and attained the Divine Vision, so that all things are made plain to him. As they talk together Asal describes the benighted state of the dwellers on the other island, and Haiy is so moved with pity at his recital that he goes over to that other island and tries to preach the higher philosophy which he has acquired. But he soon discovers that the inhabitants there are unable to rise to his teaching, and in the end came to the conclusion that their conventional religion was that best adapted to their capacity. He went back to his former home and there devoted himself to a life of solitary contemplation. This led to the conclusion that religion, as commonly accepted, following the faith revealed through Muhammad and the precepts laid down by him, is that most suitable for average humanity : speculative philosophy should be restricted to the select few who ought not to publish their conclusions to the unenlightened multitude.

NOTES

Note 1, page 8. Aramaic.

The Aramaean people were an outlying northern branch of the Arabs, nomads of the desert between Mesopotamia and Syria. They appear already in the Babylonian-Assyrian inscriptions of the fourteenth century B.C. as *Arime* or *Akhlame* and menaced the western borders of the empires of the Euphrates-Tigris valley. They invaded Syria, where there already existed a non-Semitic civilization. That civilization they adopted and developed, but imposed their own language on the older population. In course of time their language, Aramaic, replaced Assyrian in the Assyrian Empire, and finally became the lingua franca of Western Asia under the Persians, entirely replacing the older dialects of Canaan, and even spreading across to Egypt. The oldest extant documents in Aramaic are Jewish, the Aramaic portions of Ezra (4.8–6.18) and Daniel (2.4b–7.28) in the Old Testament. The Aramaic text of Ezra is of an archaic form, that of Daniel is much later. Of the third century B.C. there are inscriptions from Palmyra where an Aramaean people lived under an Arab aristocracy, and of the first century B.C. from Nabataea where an Arab people used Aramaic as a literary dialect, if inscriptions can be regarded as literary.

In Christian times Aramaic appears in two dialectal forms, Western and Eastern, the former with a phonology which has resemblances with Hebrew, probably representing the vernacular of the Syrian and Palestinian littoral, whilst the Eastern remains more true to the earlier Aramaic. The Eastern form is used in the Jewish Aramaic of the Targums and Talmud (Gemara). The Aramaic of Palestine, which gave way before the Arab conquest, is known to us only in fragments recovered of recent years from Sinai, Egypt, and Damascus. In the hinterland Aramaic survived in the western dialect only in some communities in the Lebanon, but the eastern dialect spread from the highlands of Armenia to the Persian Gulf and produced a rich literature. The focus of that literary output was at Edessa, and the material produced belongs chiefly to the Christian era, though there was a certain pre-Christian Edessene literature. But most of its material dates from the third century A.D. onwards. The Christian Aramaic writers introduced the term *Suraye* as the name of their language, a name based on the fact that its home was in the Roman province of Syria, and from that it is usual to employ the term Syriac to denote Christian Aramaic. A distinctive feature of this Aramaic is the use of the prefix *n-* in the 3rd person of the imperfect tense of the verb in place of the *y-* which appears in other Semitic languages.

Note 2, page 11. The Zoroastrian Religion.

The primitive religion of the Medes and Persians was of the Aryan type. Zoroaster was a reformer who preached probably in Media (East Persia) in the sixth century B.C. (Thus A. J. Jackson, *Zoroaster the Prophet of Ancient Iran*, New York, 1899.) No reference to him occurs in Herodotus, who refers to the Magi or members of the priestly caste and reckons them as one of the six tribes into which the Medes were divided (Herodotus, i, 101). The office of the Persian priests was not to sacrifice but to be present when sacrifice was offered and recite the proper liturgical formulae without which no sacrifice was valid (Herdt., i, 132). In addition to this exclusive knowledge of the liturgical forms the Magi were supposed to possess the power of interpreting dreams (Herdt., i, 107). Herodotus points out a striking difference between the Egyptian priests and these Magi in that the former were careful to avoid taking life except in offering sacrifice, whilst the Magi were under no such prohibition, but were ready to kill any animals, except only dogs and men (Herdt., i, 140). The Persian dead were not buried unless their bodies were first torn by a dog or some bird of prey (ibid.). The religion of the Medes and Persians had no idols, no temples or altars, but sacrifice was offered upon lofty mountains to the universe, to the sun, and moon, and to earth, fire, water, and the winds (Herdt., i, 131).

This religion, as described by Herodotus, seems to have been that of the Medes amongst whom Zoroaster preached. It was probably about the same time the Medes conquered the Persians and introduced the religious reforms of Zoroaster at least amongst the ruling Persian aristocracy. It is doubtful whether the Achaemenid kings of ancient Persia before the time of Alexander were actually Zoroastrians, but J. H. Moulton's *Early Zoroastrianism* makes a good case in favour of their being so.

The tradition is that the sacred books of the Persians were destroyed by Alexander, but it is more probable that the liturgical forms were not yet reduced to writing. Admittedly those forms exist only in fragmentary form.

When the Parthians established an independent kingdom about 238 B.C. they adopted the Zoroastrian religion and the " everlasting fire " was cherished and reverenced in the royal city of Asaak, at least until the later Parthian monarchs. Such fragments of the sacred Avesta as could be recovered were then translated into Pehlewi, which is a later form of the language used in the Avesta and inscriptions. The older language was written in cuneiform, but Pehlewi used an alphabet of Aramaic origin. The later Arsacid kings seem to have been devoted to the Zoroastrian religion until just towards the end when, it is said, the sacred fire was allowed to go out.

Apparently Zoroastrianism had several rivals, survivals of the older religion which were only partially touched by Zoroaster's reforms. It was the task of the earlier Sasanids to impose the Zoroastrian religion and to exterminate those variants as heresies. The text of the Avesta was revised and completed by a priest named Aturpat-i-Maraspandan during the reign of Shapur II (A.D. 309-379). In 456 Yezdgird II forced Zoroastrianism on Armenia where, however, it did not take hold permanently. The golden age of Zoroastrianism and that of Pehlewi literature was the reign of Khusraw I (A.D. 531-578), and at that time it was still a missionary religion which the Persian monarchs imposed on the lands they conquered. It thus spread eastwards as a rival of Buddhism without, however, exterminating the followers of Buddha. At that time Buddhism was losing ground in Central Asia, but making substantial progress in the Far East.

Note 3, page 52. Nestorius.

According to Socrates (*Eccles. Hist.*, vii, 29) there were two candidates for the see of Constantinople at the death of Sisinnius. One of these was Philip of Side who is described as an ambitious writer, the author of a work which he called not an Ecclesiastical History but a " Christian History " (Socrates, *Eccles. Hist.*, vii, 23), and the other was Proclus whom Sisennius had ordained Bishop of Cyzicum, but the people of that city refused to accept him as their bishop (ibid., 28). " At the death of Sisennius, on account of the factions and rivalries of the church as to the episcopate, it seemed good to the emperors to appoint neither, for many strove for Philip, many for Proclus, to be ordained. Therefore they decided to invite one from Antioch, for there was there a certain man, Nestorius by name, called ' the Germanican ', a good speaker and eloquent " (ibid., 29, 1-3). This makes it clear that from the beginning of his episcopate Nestorius had two sets of opponents to face.

" Nestorius brought with him from Antioch a presbyter named Anastasius," and he " preaching one day in the church said, ' Let no one call Mary the Mother of God (*theotokos*), for Mary was but a woman, and it is impossible that God should be born of a woman ' " (ibid., 32, 2-3). At that time, following the Nicene Council, the accepted doctrine was that Christ had two natures, the human and the divine, both united in one person, and Anastasius apparently intended to say that the Blessed Virgin Mary was the mother of the human nature only. But popular opinion at Constantinople at once represented Anastasius as reviving the teaching of Paul of Samosata and Photinus that Christ was merely a man. Socrates, who treats Nestorius with respect and some degree of sympathy, says that he did not hold that view nor did he deny the deity of Christ, " but he feared the term alone as though (it were) a ghost and he was alarmed at this because of great ignorance " (ibid., 32, 12). " The term " of course means " Mother of God ". It seems a logical deduction from the doctrine that Christ was God and man at his birth to give the name of Theotokos to the Virgin Mother, and the term is used by Eusebius (*De Vita Constant.*, iii, 43), by St. Cyril of Jerusalem

(*Catech.*, x, 146), and St. Athanasius (*Orat. III c. Arianos*, xv, 33), and so must have been regarded as consistent with Nicene doctrine. Hesychius, a presbyter of Jerusalem who died in 343, goes further and calls David the ancestor of Christ "father of God" (*Theopator*, Photius, *Cod.* 275). Nestorius' own explanation of his objection to the term is given by Evagrius (*Eccles. Hist.*, i, 7) : "he asserts that he was driven to assume this position because of absolute necessity because of the division of the church into two parties, one holding that Mary ought to be called Mother of Man, the other Mother of God, and he introduced the term Mother of Christ in order, as he says, that error might not be incurred by adopting either extreme, either a term which too closely united immortal essence with humanity or one which whilst admitting one of the two natures, made no reference to the other."

At the Council of Ephesus the charge was brought against Nestorius that he had stated in a discourse that "the creature did not give birth to the uncreated but bore a man, the instrument of the Deity. The Holy Spirit did not create God the Word, but made for God the Word a temple which he might occupy, from the Virgin. . . . He who was born and needed time to be formed and was carried the necessary months in the womb, had a human nature, but a nature joined with God" (Mansi, *Concilia*, iv, 1197).

The usual view of Nestorius' teaching was that Christ's body was conceived miraculously by the Holy Spirit in the Blessed Virgin Mary, but that he was born a man : the Holy Spirit afterwards descended on Him and then the Godhead entered into Him. Such is the account given by St. Augustine (*De Haeresibus*, Appendix, ch. 91). In favour of this must be cited Nestorius' words as reported by Socrates (*Eccles. Hist.*, vii, 34, 4) : "I, said Nestorius, will not call him God when He was two or three months old."

According to the teaching of Muhammad, a Spirit came from God to tell Mary that she should bear a son (Qur., 19, 19), she being then a virgin (ibid., 20), but she conceived without detriment to her virginity (ibid., 28–9). The miraculous virgin birth is asserted, but it is denied that He who was born of her was the Son of God (ibid., 36, 4, 169). The Holy Spirit was given to Him (Qur., 5, 109). His birth is treated as an act of creation : the Virgin Mother said, "How, O my Lord, shall I have a son when no man has touched me ? He said : Thus, God will create what He will : when he decreeth a thing He only saith, Be, and it is" (Qur., 8, 42) : He is as Adam, created from the dust (Qur., 19, 17–22 ; 5, 110).

Note 4, page 66. Hira.

Hira (Syriac Ḥirta) was founded about A.D. 240. It is mentioned as a Parthian town under the name of *Ertha* in Glaucus, *Fragmenta*, ed. Mullar, p. 409, and Stephanus of Byzantium, *Ethnica*, ed. Meineke, p. 276. The city consisted of a number of fortified dwellings of the kind known as qaṣr, plur. quṣūr, each a rectangle surrounding a courtyard, the enclosing wall having only one door which opened into the courtyard. The upper part of this wall had loopholes for defence and there was a bastion or tower at each corner. All the qusur were assembled round an open space which had no separate defences. There was no city wall surrounding the group, nor was there any central stronghold or citadel in which valuables might be stored. Thus when Khaled ibn al-Walid in the autumn of 634 attacked Hira the inhabitants retired to their fortified qusur which Khaled was unable to take, but they could not bring their herds or sheep into safety, but had to leave them outside. The Arabs drove off the animals and turned them into the standing harvest, at which the people of Hira asked for terms and surrendered.

The Arab population of Hira lived under the rule of the royal dynasty of the Lakhmids, whose chief was given the title of "king" by the Persian monarch. These Arabs were early in touch with Christian missionaries and a church existed there from the beginning of the fifth century. Amongst the signatures of the Council of Seleucia in 410 is that of Hosha', Bishop of Hirta'. This council is erroneously described by Musil as "Nestorian". The Nestorians did not come into existence until 430, but there were councils in the Persian Church before then. For some considerable time, however, the ruling dynasty and many of the

Arab citizens remained pagans. It was only in the days of the Patriarch Isho'yahb (582–595) that King Nu'man V was baptized by the Bishop of Hira Simeon. Nu'man's sister Hind founded the monastery called after her name Der Beni Hind, north of Hira, and there Isho'yahb's body was brought after his death at Beth Qush and buried. Isho'yahb died in exile as he had fled from Persia to escape the anger of King Khusraw. After the capture of Hira by Khaled in 634 the ruling Arabs were ordered to choose between three alternatives : (i) to embrace Islam, (ii) to pay the poll tax, or (iii) to continue war. These demands were made because the Arabs of Hira were regarded as people of Arabia for whom membership of the Muslim confraternity was compulsory. The conditions did not affect the Aramaic subject population. The Arabs of Hira consented to accept Islam, as indeed they had already done before the death of Muhammad, but had afterwards fallen away, whilst the subject population remained Christian of the Nestorian Church and became liable to the poll tax.

In the centre of Hira was another large monastery known as that of the Son of Maz'uq, and that was frequented as a pleasure resort by the people on festivals (Ash-Shabushti, *Diyaret*, MS. fo. 101r, cited by Musil, *The Middle Euphrates*, 103).

Hira appears in church history as a stronghold of Nestorianism, but it had not always been so. According to al-Ya'qubi, *Ta'rih*, ed. Houtsma, i, 258, the Iyad tribe moved from al.yemama to Hira, where they already possessed several of the qusur, but later was transferred by Kisra' (Khusraw?) to Tekrit, the central market of Upper Mesopotamia. Tekrit was strongly Monophysite and that presumably was the religious affiliation of the Iyad so, if they were Christians at the time of their sojourn in Hira, they must have given the place an anti-Nestorian tone. It is, however, very probable that they had not yet embraced Christianity when they were sojourners in Hira, nor is it at all clear that Hira was as yet Christian at that time.

Though a great Nestorian centre Hira had no Nestorian academy and Christians desiring a higher education went to Jundi-Shapur, as Hunayn ibn Ishaq did. From Ibn Masawaih's contemptuous reference to Hira and its people it seems to have been regarded as a place wholly devoted to commerce and neglectful of scholarship.

The royal court of the Lakhmids at Hira brought a tone of luxury and pomp amongst the Arabs which is reflected in the poetry of those early poets associated with Hira. The older type of " desert " poet sang about the hardships of desert life and tribal wars, mingling his song with praise of his patrons and derision of their enemies. Those poets known to have been associated with the court of Hira introduced an erotic element and often sang in praise of wine and drinking parties, subjects unfamiliar to the true desert poet. Such was not the case, however, with the poet Tarafa ibn al-'Abd, who was connected with the court of King 'Amr ibn Hind (*circ.* 554–568), because his poems were composed before he went to court. Nor was it the case with Labid ibn Rabi'a Abu 'Aqil (d. 661, 662, or 663) who boasts of being a member of the *majlis* or senate of Hira, and whose poetry shows a grave and moral element which may reflect the influence of pre-Islamic Christian teaching, a tone apparent also in the poetry of Nabigha and in that of Zuhayr, both favourites of King Nu'man ibn Mundhir of Hira. The poetry of A'sha Maymun ibn Qays contains passages which may show the influence of Christian teaching, but other passages dwell on wine and wine parties either, or both, of which may be coloured by the poet's intercourse with the Christian wine merchants of Hira with whom he dealt.

The camp-city of Kufa was founded near Hira soon after the year 638 and when 'Ali came there in 657 it already was a considerable town. As it grew the population of Hira tended to drift over to it. But the two great palaces of as-Sadir and al-Khawarnaq close by still remained in partial use, and the latter sometimes served as a hunting lodge for the earlier 'Abbasid khalifs. Hira is now represented by a mound of ruins south-east of the mound of al-Knedre, half-way between the ruins of Kufa and al-Khawarnaq (cf. Musil, *The Middle Euphrates*, p. 35, n. 26).

Note 5, page 74. Eutyches.

Eutyches was examined and condemned by a local synod held by the Patriarch Flavian of Constantinople. The proceedings of that synod are given in the Acts of the Council of Chalcedon (Mansi, *Concilia*, vi, 649 *sqq.*).   When asked to acknowledge that there were two natures in Christ he refused to do so and for this was condemned (cf. Eutyches' letter to Pope Leo in Mansi, v, 1015, " expetebar duas naturas fateri at anathematizare eos qui hoc negari ").   He supposed that the human nature was entirely absorbed in the divine.   This was the teaching attributed to the Monophysites, as their name implies, those who refused to accept the decrees of Chalcedon.   The difficulty is that those anti-Chalcedonians included several diverse groups and it was only one such group, that led by Julian, Bishop of Halicarnassus, which pressed this to a logical conclusion.   The Julianists were described as Aphthartodoketai or Phantasiastae, those who held that the human body of Christ was so infused by the Deity that it had only the appearance of humanity and was not subject to corruption, a doctrine denied by the more moderate party led by Severus of Antioch.   Both Severians and Julianists split into sub-divisions, which does not concern us at present, and ultimately the Julianists disappeared altogether, but modern works on theology commonly attribute to all " Monophysites " the doctrines of the extreme Julianists

Note 6, page 91.   Tekrit (Tagrit).

Tagrit was about thirty miles north of Samarra on the right bank of the Tigri and had a strong castle overlooking the river.   The Iyad tribe which had been removed there from Hira by Kisra' (Khusraw ?) had originally come from al-Yemama.   Tekrit was a central market for all the nomadic tribes dwelling between the Tigris and Euphrates.

In the tenth century Ibn Hawqal noted that most of its inhabitants were Christians and that there was a great monastery there.   These Christians of Tagrit were strongly anti-Nestorian and resisted Barsauma's attempt to convert them to Nestorianism in 449 (Bar Hebraeus, *Chron. Eccles.*, ii, 67, 85).   With the rise of Monophysitism they became ardent supporters of the Monophysite Church.   The chief prelate of the Persian Monophysites bore the title of Bishop of Tekrit, but for some time these prelates resided in the monastery of Mar Mattai, this for security as Monophysitism was not formally tolerated in Persia, but afterwards removed to the city of Tekrit.   The first bishop to bear the title of Mafrianus was Maruta (629).   There were twelve bishops under the Mafrianus of Tekrit as metropolitan.   When the Muslim Arabs took Tekrit in 637 Maruta surrendered the castle to them.   In the castle he built a cathedral which remained the principal church of the Persian Monophysites.   Barjesu, who was Mafrianus from 669 to 683, built a church at Tekrit in honour of St. Sergius and St. Bacchus, and later on this was recognized as a second cathedral.   Denha, who was Mafrianus after 684, consecrated bishops without the consent of the Patriarch Julian and for this was deposed and imprisoned in a monastery, but at Julian's death he was restored.   He built a church in honour of St. Ahudemmeh who had suffered martyrdom for baptizing a son of the Persian king, and this church was reckoned as a third cathedral.   In addition to these cathedrals there were several ancient and important monasteries in Tekrit.   The Mafrianus or supreme head of the Persian Monophysites ceased to reside in Tekrit after 1513.

Note 7, page 122.

Sanskrit was developed as a sacred language.   The results of this development were summed up in Panini's *Aṣṭadhyayi* probably in the fourth century B.C.   It is artificial in form, and some have supposed that it was an artificial creation designed to counteract the influence of Pali literature by recasting Prakritic language with the help of Vedic forms, but this is doubtful.   Changes took place in Sanskrit in the course of its prolonged literary history, and much of what Panini teaches is not represented in literature.   *Prakrit* is an artificial literary dialect derived from older Sanskrit.   It exists in three forms :—

(i) Primary Prakrit, of which both Vedic and Sanskrit are literary forms.
(ii) Secondary Prakrit, which includes the Prakrit of the grammarians and Pali represented in literary form by speeches, sayings, poems, tales, rules of conduct, etc., and in larger collections known as pitaka.   The Buddhist canon

consists of three such collections (tipitaka) which were finally fixed in Ceylon in the first century A.D.

(iii) Tertiary Prakrit, which is the source from which modern dialects are derived.

Note 8, page 148. al-Anbar.

Al-Anbar " the Granaries " was on the left bank of the Euphrates and was one of the greater cities of 'Iraq. It controlled an important crossing of the Tigris and was the starting-point of the trade route across the desert to Syria. The city had been founded by Shapur I who named it Buzurg (or Peroz) Shabur, and is to be identified with the Pirisuboras of Ammianus Marcellinus 24.2.9.22. It was also known as Abbareon and it was by it that the young prince Khusraw II passed on his way to seek help from the Roman Emperor Maurice.

Towards the end of the fourth century the hermit Mar Yunan made his abode in the desolate environs of the city and there he died. A church was erected over his grave, but his body was afterwards removed to the principal church in the city. Outside the city precincts was the monastery of Mar Yunan, known as the Der al-Ghurab, to which the citizens went out annually as to a pleasure resort (Abu l-Fada'il, ed. Juynboll, i, 141). This monastery was founded by 'Al al-Masih about 540, and was demolished by the Khalif al-Mutawakkil in 853. The Christians of al-Anbar or Peroz Shabur were Nestorians and their Bishop Moshe' took part in the Nestorian synod of 486 (J.-B. Chabot, *Synodicon*, 53). There was, however, also a Monophysite Bishop Aha in 629 (Michael the Syrian, *Chronicle*, ed. Chabot, iv, 413). About 600 Rabban Aphni-Maran founded the monastery (or castle) of az-Za'faran on or near a high mountain, Jebal Judi, close by Peroz-Shabur. The name az-Za'faran was given it by the Arabs, its earlier name was " the Monastery of Aphni-Maran of Khurkma ".

The first 'Abbasid khalif, Abu l-'Abbas, after his installation in the great mosque of Kufa, went to al-Anbar and made his residence, and there he died in 754. His brother and successor al-Mansur lived there until he removed to his new capital Baghdad. In 797 Narun ar-Rashid stayed in the town and found that many Persians from Khurasan had taken up their abode there. He visited al-Anbar again in 803 on his return from pilgrimage, residing in the al-'Umr mosque which was adjacent to the monastery of Mar Yunan, and whilst there had the wazir Ja'far ibn Yahya the Barmakid murdered.

Note 9, page 148. Jewish Agency.

The Jews were prominent in spreading Arabic science, especially medicine, to Egypt and the West, North Africa and Spain, beginning with Ishaq ibn Amran al-Isra'eli, who served at the court of Ziyadet Allah III (902-3) at Qairawan, partly as court physician, partly as a kind of lecturer on philosophy. He had been trained in Baghdad and was in touch with the work done there in translation and exposition of the Greek authorities. As a lecturer he was a failure because Ziyadet Allah was so given to pleasure and amusements that he had no attention to spare for philosophy. Disappointed at this Ishaq devoted himself to the further study of Greek medicine and became a pioneer in introducing it to Africa, whence it spread westwards to the Maghrab and then to Andalus. His treatise, *Kitab al-bawl*, on urine is the best medieval work on the subject. His " Guide to Physicians ", of which the Arabic text is now lost, was translated into Hebrew as *Manhig* (or *Musar*) *ha-rofe'in*, and became a favourite manual for Jewish physicians. He seems to have been the first Arabic medical authority introduced to the Christian west in a Latin translation by Constantine the African (1087), which was afterwards printed at Leiden in 1515. From his time onwards Jewish physicians, then astronomers and philosophers, played a prominent part in transmitting to the west Greek science as known and interpreted in Baghdad.

But before Ishaq there were Jewish physicians in Egypt and Syria, though there are no details of their activities. Presumably they were in touch with the renaissance of Greek science which stirred the Hellenistic world and had its repercussion in the Aramaic (Syriac) community, and perhaps the Jews had an independent transmission from Alexandria which was a great Jewish centre.

The medical writer Abu l-Hasan 'Ali ibn Sahl ibn Rabban (d. 850) was a Muslim but the son of a Jewish physician of Marw, and was the teacher of Muhammad ibn Zakariya ar-Razi (Rhazes or Rases), so obviously Greek medical science had already reached the Jews of Eastern Persia. Mashallah ibn Athari (d. 815–820), one of the astrologers called in by al-Mansur at the foundation of Baghdad, is said to have been a Jew. Our general conclusion must be that there were Jewish scientists, and especially physicians, in touch with the revival of Greek science which was in progress during the eighth century, though none of these seem to have been of great prominence before Sahl ibn Rabban and Ishaq ibn Amran.

Was there any independent Hellenistic revival amongst the Jews? It does not appear that such was the case. There was a succession of Jewish teachers and schools from the last days of Jerusalem onwards, but these were concerned with the law of Moses and traditions illustrating and explaining the law. Under the Sassanids there were distinguished rabbinical schools at Nehardea on the Nehar between the Tigris and Euphrates, at Machusa on the Tigris near Ctesiphon, at Sora on the Euphrates about 20 parasangs from Nehardea, and at Pumbaditha. These had a somewhat chequered history, but under Khusraw II they prospered and are said to have included scientific research as well as purely rabbinical studies in their work. How far this actually was the case is not clear. Samuel of Nehardea (d. 250) is said to have been learned in astronomy, but at that early date when scientific material was accessible only in Greek it probably did not amount to much. Most likely it meant the computation of dates, festivals, and times of fasting, parallel with the computation of Easter which passed as astronomy amongst Christians. The fuller development of scientific studies seem to have come much later and to have been due to contact with the Syriac world which had adopted Greek science in an Aramaic version, and to have reached maturity about the time of the foundation of Baghdad, or a little later under Harun ar-Rashid. It appears that Sa'da Gaon at Pithom (al-Fayyum) in Egypt (892–942) who made translations from Hebrew into Arabic was mainly responsible for making Arabic replace Hebrew or Aramaic as the literary language of Judaism, and as long as this use of Arabic continued the Jews were in close contact with contemporary Arabic scientific and philosophical thought. When the use of Hebrew was revived translations were made from Arabic into Hebrew, and many Arabic scientific works are now known to us only in these Hebrew versions. A survey of this material shows that Jewish interest was most prominent in medical studies. The Jews played a leading part in transmitting scientific material from Arabic to Latin, chiefly through Cordova, Toledo, and Barcelona. Earlier Latin versions connect with Monte Cassino, Tyre, and (Syrian) Tripoli, later with the Dominican friars in Syria, and these were not indebted to Jewish workers, though they seem to have selected Jewish works such as these of Ishaq ibn Amran as best suited for teaching medical science to the Christian west.

BIBLIOGRAPHY

ABU-L-FEDA. *Annales Muslemici*, Arab.-lat., 5 vols., Copenhagen, 1789-94.
AHUDEMMEH. " Life," ed. F. Nau in *PO.*, iii, fasc. 1, Paris, 1906.
ALLMAN, G. J. *Greek Geometry from Thales to Euclid*, Dublin, 1889.
AMMIANUS MARCELLINUS. Tauchnitz edit., Leipzig, 1876.
ARNOLD, T. W. *Preaching of Islam*, 2nd edit., London, 1913.
—— *The Caliphate*, London, 1924.
ASSEMANI, J. S. *Bibliotheca Orientalis*, i-iii, Rome, 1719-1728.
BAR HEBRAEUS. *Chronicon Ecclesiasticum*, ed. J. B. Abbeloos et T. J. Lamy, Louvain,
   1872-7.
—— *Chronicon Syriacum*, ed. P. Bedjan, Paris, 1890.
BAUMSTARK, A. *Geschichte der syrischen Literatur*, Bonn, 1922.
EL-BELADHURI. *Kitab futuh al-buldan, Liber expugnationis regionum*, ed. J. de Goeje,
   Leiden, 1868.
BERGESTRÄSSER, G. *Risalat Hunayn ibn Ishaq*, Leipzig, 1925.   (Analysis by
   Meyerhof in *Isis*, viii (1926), 685-724.)
BEVAN, E. R. *House of Seleucus*, 2 vols., London, 1902.
—— *Hellenism and Christianity*, London, 1921.
DE BOER, T. J. *Geschichte der Philosophie im Islam*, Stuttgart, 1901. (Inadequate
   but the best available.)
BOUYGES, A. M. *Sur le de scientiis d'Alfarabi*, Beyrouth, 1924.
BROCKELMANN, C. *Geschichte d. arabisch. Literatur*, 2 vols. I. Weimar, 1898 ;
   II. Berlin, 1902. Supplementary fasciciles, 1937, etc. (Chiefly biblio-
   graphy. An indispensable work of reference, but with occasional
   inaccuracies.)
BROOKS, E. W. " Vitae virorum apud Monophysitas celeberrimorum " in
   CSCO., ii, 25. Paris, 1907.
BROWNE, E. G. *History of Arabian Medicine*, Cambridge, 1921.
—— *Chahar Maqala*, 2 vols., London, 1910.
—— *A Literary History of Persia*, New York, 1902. (Introductory part gives a
   general account to the early cultural history of Islam.)
CAETANI, L. *Annali dell' Islam*, vols. i, ii., Milano, 1905-7. (Best account of
   rise and spread of Islam, but in several details corrected by Musil q.v.)
CAJORI, F. *A History of Mathematics*, New York, 1924.
Cambridge History of India, vol. i, Cambridge, 1922.
CANTOR, M. *Vorlesungen über Gesch. der Mathematik*, Leipzig, 1907.
CARRA DE VAUX. *Penseurs d'Islam*, 5 vols., Paris, 1921-8.
—— *Avicenne*, Paris, 1900.
—— *Mas'udi, le livre de l'Avertisement, trad.*, Paris, 1897.
CHABOT, J-B. " L'École de Nisibe " in *JA.*, 1896.
—— " Documenta ad origines Monophysitarum illustrandas " in *CSCO.*, ser. ii,
   vol. 37, Paris, 1903.
—— " Synodicon orientale " in *Notices et extraits*, xxxvii, Paris, 1902.
CHRISTENSEN. " L'empire des Sasanids " in *Jour. Iran. Assoc.*, viii, 434.
" Chronicle of Edessa " in *Texte und Untersuch.*, IX, i, Leipzig, 1893.
CHWOLSON, D. *Die Ssabier und der Ssabismus*, 2 vols., St. Petersburg, 1856.
CRUM, W. E. " Sévère d'Antioche en Égypte " in *Rev. Orient. Chrét.*, iii (192-3),
   92-104.
CSCO., *Corpus Scriptorum Christianorum Orientalium*, Paris.
CUMONT. *L'Égypte des astrologues*, Bruxelles, 1937.
DARMESTSTER. " Lettre de Tansar au roi de Tabaristan " in *JA.*, 144, 186.
   (Shows that neo-Platonism had penetrated into Persia.)
DAVIES, R. *Buddhist India*, London, 1903.
DENHA. " Histoire de Marouta " in *PO.*, III, 52-96.
DIEHL. *Justinien*, Paris, 1901.
DIETERICI, F. *Alfarabi's philosophische Abhandlungen*, Leiden, 1890.
DOUGHTY, C. M. *Travels in Arabia Deserta*, 2 vols., London, 1923.

DREYER, J. L. E. *History of the Planetary Systems*, Cambridge, 1903.
DROYSEN, J. G. *Gesch. de Hellenismus*, 3 vols., 2nd edit., Gotha, 1877–9.
DUCHESNE, L. *Early History of the Christian Church*, Eng. trs. of 4th edit., 3 vols., London, 1914.
—— *Églises separées*, Paris, 1906.
—— *L'église au vi^e siècle*, Paris, 1929.
DULSEM, P. *La système du monde*, Paris, 1915.
"Elias of Nisibis, Opus chronologicum," ed. E. W. Brooks and J-B. Chabot, in *CSCO.*, iii, vols. 7, 8, Paris, 1909–11.
Encyclopaedia of Islam, ed. T. Houtsma and others, Leiden, 1906–34. Supplement, 1938.
EVAGRIUS. "Historia Ecclesiastica " in *PG.*, lxxxvi, 2415 *sqq.*
FLÜGEL, G. *Al-Kindi, genannt " der Philosoph der Araber "*, Leipzig, 1857.
—— "Ueber Inhalt und Verfasser der arabischen Encyclopädie der Ikhwan as-Safa " in *ZDMG.*, xiii, 1 sqq.
GOLDZIHER, J. *Muhammedanische Studien*, 2 vols., Halle, 1889–90.
GOODSPEED. "Athanasius (of Antioch), Conflict of Severus " in *PO.*, iv, 333–590.
Hamza al-Isfahani, ed. J. M. E. Gottwaldt, St. Petersburg, 1844.
HANKEL, H. *Zur Geschichte der Mathematik*, Leipzig, 1874.
HARNACK, A. *Lehrbuch der Dogmengeschichte*, 3 vols., Freiburg, 1894.
—— *Geschichte der altchristlichen Litteratur*, Leipzig, 2 vols., 1893.
—— *Die Chronologie des altchristlichen Litteratur*, Leipzig, 2 vols., 1897, 1904.
HASKINS, C. H. " Arabic Science in Western Europe " in *Isis*, vii (1925), 478–486.
—— *Studies in the History of Medieval Science*, Camb., U.S.A., 1924. (Good account of Latin versions of Arabic scientific works.)
HAUSER. *Ueber das Kitab al-hijar*, Erlangen, 1922. (Account of the " Sons of Musa ", etc.)
HEATH, T. L. *Aristarchus of Samos*, Oxford, 1913.
—— *History of Greek Mathematics*, 2 vols., Oxford, 1921.
HEFELE, C. J. *History of the Christian Church Councils*, English transl., 4 vols., Edinburgh, 1871–83.
HEURTLEY, C. A. *De fide et symbolo*, Oxford, 1887.
HIRSCHBERG, J. *Geschichte d. Augenheilkunde*, Leipzig, 1899–1918.
HOERNLE, A. F. R. " Studies in Ancient Indian Medicine " in *JRAS.* (1906), 233–302, 915–941 ; (1907), 1–13 ; (1908), 997–1028.
HOFFMANN, J. G. E. *De Hermeneuticis apud Syros Aristotelis* (Syriac), Leipzig, 1873.
HOGARTH, D. G. *The Nearer East*, London, 1905.
HOMMEL, F. *Grundriss der Geogr. u. Gesch. des altens Orients*, i, 1904 ; ii, 1926.
HUART, C. *Histoire des Arabes*, Paris, 1911–12.
Ibn Abi Usaibi'a, ed. A. Müller, 1884. (Biographies of eminent physicians.)
Ibn Khallikan, Wafayat al-a'yar wa-anba' abna' az-seman. Edit. Wüstenfeld, Götingen, 1836–71. English trans., Baron MacGluckin de Slane, Paris-London, 1842–71. (Biographical dictionary finished in 1274.)
INGE, W. R. *Philosophy of Plotinus*, London, 1918.
INOSTRANZEV. *Iranian influences on Moslem literature*, Bombay, 1918.
IORGA, N. *Rélations entre l'Orient et l'Occident au moyen âge*, Paris, 1923.
Isis, periodical dealing with history of science : ed. G. Sarton.
JA., *Journal Asiatique*, periodical, Paris.
Janus, "Zeitschrift für Geschichte und Litt. des Medizin," Leiden, 1924.
JOHN OF APHTHONIA. *Life of Severus*, ed. trs. M. A. Kugener, in *PO.*, II, iii, Paris, 1905.
JOHN DAMASCENE. In *Migne Patrologia Graeca*, xciv and xcvi.
JOSHUA THE STYLITE. *The Chronicle of Joshua the Stylite*, ed. W. Wright, Cambridge, 1882.
JRAS., *Journal of the Royal Asiatic Society*, periodical, London.
KARPINSKI, L. C. *Robert of Chester's Latin Translation of the Algebra of al-Khwarizmi*, New York, 1915.
AL-KINDI. *Defence of Christianity*, Engl. trs. Sir William Muir, *The Apology of al Kindy, with an essay on its age and authorship*, London, 1911. (The work of a Nestorian monk under al-Ma'mun.)

KING, L. W., and THOMPSON, H. R. *Sculptures and Inscriptions of Darius the Great on the Rock of Behistan*, London, 1907.
KOHL, K. " Ueber den Aufbau der Welt nach Ibn al-Haitham " in *Sitzb. d. phys. med. Soc.*, Erlangen, 1925.
VON KREMER, A. *Culturgeschichte Streifzüge auf dem Geniete des Islam*, Leipzig, 1873.
—— *Culturgeschichte des Orients unter den Chalifen*, Wien, 1875-7.
—— *Geschichte der herrschen Ideen des Islam*, Leipzig, 1868.
LABOURT, J. *Le Christianisme dans l'Empire Perse*, Paris, 1904.
LAMMENS, H. *Le Chantre des Omiades*, Paris, 1895.
—— *La Mecque à la veille de l'Hégire*, Beyrouth, 1924.
—— *L'Arabie occidentale avant l'Hégire*, Beyrouth, 1928.
—— *Études sur le régne du Calife Omayade Mo'awia*, 1er, Beyrouth, 1906, 1908.
—— *Le Califat de Yazid*, 1er, Beyrouth, 1909–21.
LAND, J. P. N. *Anecdota Syriaca*, Leiden, 1862.
LANDBERG, GRAF VON. *Études*, Leipzig, 1909.
LANE-POOLE, S. *The Mohammedan Dynasties*, London, 1895.
—— *Studies in a Mosque*, 2nd edit. London, 1893.
LECLERC, L. *Histoire de la medicine arabe*, 2 vols., Paris, 1876.
LE STRANGE E. *Palestine under the Moslems* (550–1500), London, 1890.
—— *Baghdad*, 2nd edit., Oxford, 1924.
—— *Lands of the Eastern Khalifate*, Cambridge, 1909.
VON LIPPMANN, E. C. *Gesch. der Zuckers seit der ältesten Zeiten*, 2nd edit., Berlin, 1929. (Use and spread of sugar cane as indicating culture drift.)
LOEW. *Aramaische Pflanzennamen*, 1881.
LYDE, L. W. *The Continent of Asia*, London, 1923.
MACDONALD, D. B. *Development of Muslim Theology*, London, 1903.
McCRINDLE, J. W. *Topography of Cosmas*, Hakluyt Society, 1897.
MANECKJI NUSSERVANJI DHULLA. *Zoroastrian Civilization*, New York, 1922.
MASPERO–FORTESCUE–WIET. *Histoire des patriarches d'Alexandrie depuis la mort de l'empereur Anastase jusqu'à la réconciliation des églises jacobites*, Paris 1923.
MAS'UDI. *Muruj adh-Dhab*, text trs. B. de Maynard et P. de Courteille, Paris, 1861–71.
—— *Le livre de l'avertissement*, trs. Carra de Vaux (q.v.).
MERIVALE, C. *History of the Romans under the Empire*, 8 vols., London, 1896.
MEYER, E. VON. *Gesch. der Botanik*, Leipzig, 1856.
—— *Gesch. der Chemie*, Leipzig, 1914.
MEYER-STEINEG und SUDHOFF. *K. Gesch. der Medizin*, Jena, 1922.
MEYERHOF, H. " New Light on Hunayn ibn Ishaq " in *Isis*, viii (1926), 685–724.
—— *The Book of the Ten Treatises on the Eye ascribed to Hunayn ibn Ishaq*, Cairo, 1928.
—— " An Arabic Compendium of Medico-philosophical Definitions " in *Isis*, x (1926), 340–9.
MIELI, A. *Pagine di storia della Chimica*, Roma, 1922.
MOMMSEN, T. *Provinces of the Roman Empire*, Eng. trans., vol. 2, London, 1909.
MUIR, Sir WILLIAM. *The Caliphate, its Rise, Decline, and Fall*. London, 1891.
MÜLLER, A. *Der Islam im Morgen- und Abenland*, 2 vols., Berlin, 1885-7.
—— *Die Beherrscher der Gläubigen*, Berlin, 1882.
MÜLLER, M. *Die* Quaestiones naturales *des Abelardus von Bath*, Münster, 1934.
MUSIL, A. *The Manners and Customs of the Rwala Bedounis*, New York, 1928.
—— *Arabia Deserta*, New York, 1927.
—— *Palmyrena*, New York, 1928.
—— *Northern Negd*, New York, 1928.
NALINAKSHA DUTT. *Early Monastic Buddhism*, i, Calcutta, 1941.
NAU, F. " Documents pour servir à l'histoire de l'Église Nestorienne " in *PO.*, xiii, fasc. 2, Paris.
NEUBERGER, M. *Gesch. der Medizin*, Stuttgart, 1908. Engl. trans., Oxford, 1925.
NICHOLSON, R. A. *Literary History of the Arabs*, 4th ed., London, 1922.
NÖLDEKE, TH. *Gesch. der Perser und Araber zur Zeit der Sassaniden*, Berlin, 1879.
—— *Die Ghassaniden Fürsten*, Berlin, 1887.
PAGEL. *Einfuhrung in die Gesch. der Medizin*, Berlin, 1898.
PURGITER. *Ancient Indian Historical Tradition*, 1922.
PG., Migne's *Patrologia Graeca*.

192   BIBLIOGRAPHY

PINES, S.   Beiträge zur Islamischen Atomenlehre, Berlin, 1936.
PO., Patrologia Orientalis, ed. Mgr. Gräffin, Paris.
PROCOPIUS.   Ed. Dindorf, Corpus Script. Hist. Byzant., Bonn, 1833–8.
RAY, Sir PRAPHULLA CHANDRA.   A History of Hindu Chemistry, 2nd ed., Calcutta,
       n.d.
RAYMOND, A.   Histoire des sciences exactes et naturelles, Paris, 1924.
SCHWARTZ, E.   Concilium universale Chalcedonense, Berlin, 1932.
SÉDILLOT.   Prolégomènes des tables astronomiques d'Oloug-Beg, Paris, 1853.
SEEMAN, H., und MITTELBAUM, T.   Das Kugelförmige Astrolab, Erlangen, 1925.
SEWELL, A.   " Roman Coins found in India " in JRAS. (1903), 541 sqq.
SMITH, D. E.   History of Mathematics, 2 vols., New York, 1923–5.
—— and KARPINSKI, L. C.   Hindu-Arabic Numerals, New York, 1911.
SMITH, V. A.   Early History of India, 3rd ed., Oxford, 1914.
—— Asoka, 3rd ed., Oxford, 1920.
SOCRATES.   Ecclesiastica Historia, ed. Oxford, 1844.
SOZOMAN.   Ecclesiastica Historia, ed. Migne, PG., lxvii.
STAPLETON & AZO-HUSSAIN.   Chemistry in Iraq and Persia in the Tenth Century,
       Calcutta, 1927.
STEELE, R.   " Practical Chemistry in the Twelfth Century (Rasis de aluminibus
       et salibus) " in Isis, xii (1929), 10–96.
STEINES, H.   Die Mu'taziliten oder die Freidenker im Islam, Leipzig, 1865.
STEINSCHNEIDER, M.   Die europäischen Übersetz. dem arabischen bis Mitte des xvii
       Jahrhund., Wien (Sitz. des Akad.), cxlix, 22–44.
—— Die hebräischen Uebersetzungen der Mittelalters, Berlin, 1893.
STRZYGOWSKI, J.   Der Ursprung des Christlichen Kirchenkunst, 1919. Eng. trs. Dalton,
       Origin of Christian Church Art, 1923.
SUTER, H.   Die Mathematiker und Astronomen der Araber and ihre Werke, Leipzig,
       1900–4.
—— Das Buch der geom. Konstructionen der Abu 'l-Wefa', Erlangen, 1922.
—— und WIEDEMANN, E.   " Ueber al Biruni und seine Schriften " in Sitz. d.
       Physik. mediz. Gesell. (1920), 55, 96.
AT-TABARI.   Annales, ed. M. J. de Goeje and others, Leiden, 1874–1901.
TANNERY, P.   Recherches sur l'histoire de l'astronomie ancienne, Paris, 1893.
TARN, W. W.   The Greeks in Bactria and India, Cambridge, 1938. (Very useful.)
—— Hellenistic Civilization, London, 1930.
THOMAS, J.   Selections illustrating the History of Greek Mathematics (Loeb Classical
       Library), 1941.
TROPFKE, J.   Geschichte der Elementar-Mathematik, 3 vols., Berlin, 1921–2.
WARMINGTON, E. H.   The Commerce between the Roman Empire and India, 1928.
WEINBERG, J.   Die Algebra des Abu Kamil Soga' ben Aslam, Munich, 1935.
WIBERG, J.   " The anatomy of the brain in the works of Galen and 'Ali 'Abbas "
       in Janus, xix (1914), 17–32, 84–104.
WIEDEMANN, E.   Über Thabit ibn Qurra, sein Leben und Wirken, Erlangen, 1922.
—— " Zur nabat. Landwirt-schaft von Ibn Wahschija " in Zeit. f. Semit., i (1922),
       201.
—— und FRANK, J.   Ueber die Konstruction der Schattenlinien von Tabit ibn Qurra,
       Köpenhavn, 1922.
WIELEITNER, H.   Gesch. der Mathematik, i, 1921, Berlin.
WINER, L.   Contributions towards a history of Arabico-Gothic Culture, New York,
       1917. (Arguments rash and unproved.)
WOEPCKE.   Sur l'introduction de l'arithmétique indien en occident, Paris, 1859.
WÜSTENFELD, F.   Gesch. der arab. Aertze u. Naturforscher, Göttingen, 1840.
—— Die Academien der Araber, Göttingen, 1837.
WRIGHT, W.   History of Syriac Literature, London, 1894. (Cf. also Joshua.)

INDEX

(The article *al-* etc. and *Ibn* are not reckoned in classification.)

Aaron of Alexandria, 35, 92
'Abbasid revolution, 146 sqq.
'Abd al-Malik, khalif, 136
'Abd al-Masih, 159, 177
Abraham of Kashkar (two of this name), 66
Adiabene, 90
Aedisius, 28
Aetius, 35
Agapetus, Pope, 83-4
Agatharchides, 98
Ahudemmeh of Tagrit, 90
al-Akhtal, poet, 136
Aleppo, 178
Alexander the Great, 6, 110, 121
Alexandria, in Egypt, 19; schools of —, 20; Hellenism of —, 19; curriculum of the schools, 167
Alexandria "Under the Caucasus", 110, 124 sqq.
'Ali ar-Rida, 162
Almajest, 158
Ammonius Saccas, 23, 29, 176
Anastasius, 52, 53
Andalus, 170
Andrew of Crete, 138
Antiochus IV, 114
Appollonius, 30
Aqaq (Acacius), 59
Arabic grammarians, 144
Arabs of Roman (Syrian) frontier, 134 sqq.
Aramaic, 7, 182
Archimedes, 30
Ardashir, 12
Aristarchus, 30
Aristotle, 21, 70, 151, 159, 176 sqq., 178
Armenia, 118
Armenian Church, 77
Arrian's *Indica*, 101
Arsacids, 111
Aryabhata, 104, 107
Asoka, 121 sqq.; edicts of, 122
Assassins, 180
Astrology, 4
Aurelian, 16
Aurelius, 25
Axumites, 24, 102

Babowai, 58
Bactria (Balkh), 2, 3, 110, 112, 113, 126 sqq.; holy land of Buddhism, 115, 127
Baghdad, 3, 72, 146, 161, 162; foundation of —, 148

Bait Lapat, 58
Balkh, see Bactria
Bamiyan, 130
Bar Daisan, 126
Barmakids, 119, 150
Barsauma, 56, 58 sq., 61-2
Basra, 143, 144, 145, 149
Batini doctrine, 180
al-Batriq, 159
al-Biruni, 106-7
Brahmagupta, 108
Buddhism, 3, 120 sq., 183; turns towards Far East, 114; How far did it spread west? 125.
Budh, 69, 155-6
Bukhtyishu', 149

Callistus, 41 sqq.
Calotychius, 82
Camp cities, 142
Cappadocia, cradle of Greek renaissance, 49-50
Ceylon Chronicle (Buddhist), 124
Chalcedon, Council of, 56, 75
Chinese pilgrims to Balkh, 116
Christianity spread by persecution, 40 sq.
Clement of Alexandria, 125
Constantine, 15, 17, 104
Constantine VII, 171
Cordova, khalifs of, 171
Cosmas, 139
Cyril of Alexandria, 54, 73

Dadisho', 60
Damascius, 28
Damascus, 135, 138, 146
Dar al-Hikhma, 166-7, 169
Diocles, 31
Diocletian, 15-16
Diodorus, 48, 52
Dionysius the Areopagite, 78
Diophantus, 33
Dioscorides, 169, 171
Dioscoros, 54, 72, 75, 76
Döllinger, 42
Domnus, 55
ad-Du'ali, 143 sqq.

Ecclesiastical organization, 44
Edessa, 8; school of, 50-2; school closed, 57
Egypt, Hellenization of, 7
Enneads of Plotinus, 25 sq.
Ephesus, Council of, 54, 56, 74
Ephræm of Edessa, 47

Eratosthenes, 30
Ethiopia, 94
Ethiopians in the Tihama, 94-5
Euchratides, 114
Euclid, 29, 157-8
Eudoxus, 102
Eustathius, 47, 48
Eutyches, 54, 74, 185

Fa-hien's travels, 128
al-Farabi, 178
Fatimid khalifs, 180
Flavian of Constantinople, 54
Freedom of the will, 141
Fustat, 143

Gabriel of Shissar, 93
Gabriel, see Jibra'il
Gandhara, 113-4, 116
Gordian, 13
Greek, use of, 7, 36 ; Greek original
   of Indian mathematics and astro-
   nomy, 106-7
Gupta kings, 104

Hadrian, 11
Harran, 172-5
Harsha, 128
Harun ar-Rashid, 150, 151, 155, 160-1
Henoticon, 85
Herat, see of, 91
Heron, 31
Hiung-nu, 22, 115
Hibha (Ibas) 49 sq., 55, 56
Hierocles, 28
Hierotheus, pseudo-, 79
Hijaz trade-route, 98
Himyarites, 100, 102
Hippalus, 101
Hipparchus, 31
Hippolytus, 41
Hira, 66, 67, 184
Ibn Hisham, 134
Hiuen-Tsang's travels, 129
Hubaysh b. Ishaq, 84, 154, 164 sqq.
Hypathia, 28, 23
Hypsicles, 31

Iamblichus, 22, 27
Ibrahim b. Adham, 130
Ikhwan as-Safa, 179
" Incense route ", 99
India, 2 ;— Indian medicine, 69 ;
   influence of India, 96 sq. ; early
   intercourse, 96-7 ; Indian astro-
   nomy, 104
'Isa b. Yahya, 169
Isaac of Antioch, 51
Ibn Ishaq, 134
Ishaq b. Hunayn, 169
Isho' the Stylite, 88
I-tsing's travels, 129,

Jabia, 143
Jackson, A. J., 182
Jacob of Nisibis, 47-8
Ja'far the Barmakid, 72, 109, 150, 153,
   157, 160
Jain religion, 120
Jewish Diaspora, 37 sq., — influence,.
   158
Jibra'il I, 159
Jibra'il II, 159-160, 166
Jirjis b. Bukhtyishu', 150
John of Aphthonia, 89
John bar Cursus, 88
John Damascene, 139, 164
John of Ephesus, 89-90
John of Nisibis, 62
John Philoponus, 29, 92, 176
Joseph Huzaya, 62
Judaism, reactionary element, 38-40
Julian, Emperor, 17, 27, 28
Julian of Halicarnassus, 80-2
Jundi-Shapur, 150, 151, 157, 163, 164 ;
   founded, 14 ; Mani executed there,
   16 ; in ruins, 17 ; language, 71-2 ;
   observatory at, 105, 158
Justinian, 28 sq., 82

Kalilag wa-Dimnag, 69, 155-6
Kanishka, 103, 127
Kennesrin (Qen-neshre), 87, 91
Khalid b. Barmak, 148
Khurasan, 147
Khusraw I, 68-9, 183
al-Khwarizmi, 152, 154
al-Kindi, 176
Kufa, 143, 144, 149
Kunnash, 160
Kushan, kings, 103, 115, 127 ; decline
   of, 103

Lakhmids, 184
Lammens, Henri, S. J., 95
Latrocinium, 55, 74
Leontopolis, temple at, 37
Leo Tacticus, 142
Lewis, B., 180
Longinus, 23

Macdonald, D., 179
Mafrian, head of Persian Monophy-
   sites, 91
Magadha, 120
Malwa, 104
al-Ma'mun, Khalif, 161 ; measures
   earth's arc, 163
Mani, 16
al-Mansur, Khalif, 148-9
al-Mantiqi, 170
Mari the Persian, 60
Marinus, 32
Mar Mattai monastery, 90
Mashallah, 148, 153

Marutha, 95
Marw, 2, 95, 112, 117, 150, 155, 156, 158, 160, 161
Ibn Masawaih, 163–6
Abu Ma'shar, 180
Mas'udi, 152
Mazdean religion, 11 ; reformed, 119
Medicine, 4
Megasthenes, 121, 125
Melinda, 114
Menelaus, 32
Mesopotamia a Roman province, 11
Mesopotamian Church, 45
Monophysites, 73 sqq. ; persecuted, 80, 85 ; missions, 93–4
Moulton, J. H., 183
Muhammad's designs to spread Islam, 133 sqq.
al-Muqaffa', 155
Musa, Sons of, 165–6
al-Mu'tasim, Khalif, 167
Mutawakkil, Khalif, 168 sqq.
Mu'tazilite sect, 140, 141, 145, 162, 177

Najran in Arabia, 95
Nanda dynasty, 121
Narsai, 59, 62
Nawbahar, 129
an-Nawbakht, 148, 153
Nearchus, 99
Neo-Platonism, 21, 142
Nestorians, 47 sq., 168 ; schism amongst, 64 ; reformed, 64 sq. ; missions, 67–8
Nestorius, 52–4
Nicomachus, 32
Nicomedes, 31
Nisibis, 47 sqq. ; school of, 61, 66 sq.
Numenius, 23

Oribasius, 34
Origen (two of this name), 23–4
Osrohene and Edessa, 118

Palmyra, 13, 15
Pappus, 33
Parthia founded, 111 ; its expansion checked, 117
Pataliputra, 114
Paul of Aegina, 35
Paul the Persian, 69
Paul of Tella, 91
Periplus Maris Erythrei, 100–1
Persianization of the Nestorian Church, 62, 64
Peter Mongus, 81
Phantasiasts, 81–2
Philoxenus, 87
Pliny, Natural History, 100
Pompey conquers Syria, 118
Porphyry, 22–4, 26 sqq.
Probus, 82

Proclus, 28, 34
Proterius, 75–6
Ptolemy, Claudius, 32, 101
Ptolemy Philadelphus, 20
Ptolemy Soter, 20
Pythagoras, 21 sq.

Qairawan, 143
Qamisho', 90
Qualities (sifat) of God, 141–2
Qur'an, eternity of, 140–1

Rabbula, 49 sq., 53
Raishahar academy, 70
Rakote, 19
Roman Law, 139–140
Roman power in the Near East, 8–10
Ross, Sir Denison, 1

Sabaeans, 100
Sagala, 114
Sakas, 102, 104, 115 ; Saka states, 115
Sasanids, 12 ; Sasanid revolution, 119 ; Sasanid " Royal Tables ", 105, 156
Scetis, Syrian monastery in, 92
Septuagint, 37 sq., 43
Sergius of Damascus, 138 sq.
Sergius of Rashayn, 83
Severus, Emperor, 11
Severus of Antioch, 77 sq., 84–5 ; patriarch, 78, 80 ; in Egypt, 81
Severus Sebokht, 93
Shapur II, 17
Shem'on, historian, 58, 59
Shem'on of Beth Garmai, 71
Shem'on of Beth Arsham, 88
Shem'on Quqaya, 89
Shi'ites, 3, 147
Shu'ubiya, 156
Sindhind, 105, 152
Skylax and the sea-route to India, 97, 99, 110
Sophronius, 138
Stephen bar Sudhaili, 79
Strabo, 101, 102, 125
Strzygowski, 47
Sugar, 71
Syria, Hellenization of, 6 ; a Roman province, 10 ; conquered by the Arabs, 131 sq.
Syriac, 8, cf. Aramaic

Tagrit, 91, 186
Tetrabiblos, 159
Thales, 21
Theodora Empress, 82, 85, 94
Theodore of Bostra, 86
Theodore of Marw, 66, 84
Theodore of Mopseustia, 48, 52
Theodorus Abucara, 139
Theodosius, Emperor, 75
Theodosius of Alexandria, 82, 86

Thomas of Harquel, 91
Timothy Aelurus, 76
Trajan, 10
Tur 'Abdin monastery, 91

Ujjain, 104
'Umayyad dynasty, 135 sqq.
'Umayyads of Spain, 8, 147 sq.

Valerian's defeat, 12
Varahamihisa, 105

Ya'qub Burde'ana, 85 sq.
Ya'qub of Sarug, 87
Ya'qub b. Tariq, 152
Yazdegerd I, 59–60
Yemen, route to India, 98
Yueh-chi, 102, 115

Zacharias Rhetor, 90
Zethas, 25
Ziyad b. Abihi, 144
Zoroaster, 119